ENERGY POLICIES, POLITICS AND PRICES

CHINA'S WIND AND SOLAR INDUSTRIES

ISSUES, TRENDS, AND IMPLICATIONS FOR THE U.S.

ENERGY POLICIES, POLITICS AND PRICES

Additional books in this series can be found on Nova's website under the Series tab.

Additional e-books in this series can be found on Nova's website under the e-book tab.

ENERGY POLICIES, POLITICS AND PRICES

CHINA'S WIND AND SOLAR INDUSTRIES

ISSUES, TRENDS, AND IMPLICATIONS FOR THE U.S.

RITA SCHWARTZ
EDITOR

nova publishers
New York

NOTICE TO THE READER

Library of Congress Cataloging-in-Publication Data

ISBN: 978-1-63483-166-6

Published by Nova Science Publishers, Inc. † New York

CONTENTS

Preface		**vii**
Chapter 1	China's Wind and Solar Sectors: Trends in Deployment, Manufacturing, and Energy Policy *Iacob Koch-Weser and Ethan Meick*	**1**
Chapter 2	Statement of Leocadia I. Zak, Director, U.S. Trade and Development Agency. Hearing on "U.S.-China Clean Energy Cooperation: Status, Challenges, and Opportunities"	**85**
Chapter 3	Testimony of Joanna Lewis, Assistant Professor of Science, Technology and International Affairs, Georgetown University. Hearing on "U.S.-China Clean Energy Cooperation: Status, Challenges, and Opportunities"	**97**
Chapter 4	Testimony of Sarah Forbes, Senior Associate, World Resources Institute. Hearing on "U.S.-China Clean Energy Cooperation: Status, Challenges, and Opportunities"	**105**
Index		**117**

PREFACE

Renewable energy is gaining currency around the globe, but China and the United States are central to its development. They are the world's top-two countries in terms of energy consumption, net oil imports, and carbon emissions, as well as gross domestic product (GDP) and manufacturing. Their large territories harbor some of the best sites to generate renewable energy. If the United States and China cooperate—and compete—effectively, renewable energy can contribute to economic growth, energy security, and climate change mitigation. The past two years saw important developments, including a bilateral agreement to phase down hydrofluorocarbon emissions and a joint announcement to reduce greenhouse gas emissions. China is often touted for its rapid adoption of renewable energy technologies. Long-term industry plans and new legislation suggest Beijing will maintain this policy direction, in line with its overall expansion of energy production capacity. In the United States, by contrast, renewable energy is a divisive issue. There are disagreements about how, if at all, the government should support alternative energy sources when natural gas is abundant, emissions are declining, and energy demand is slowing. This book assesses recent developments in China's wind and solar industries and the implications for the United States. It builds on the Commission's past work on U.S.-China energy issues, including the April 2014 hearing on bilateral clean energy cooperation. The research also draws on Congressional testimonies, academic papers, industry and media reports, and statistical data. The report's main themes and findings are outlined below.

In: China's Wind and Solar Industries
Editor: Rita Schwartz
ISBN: 978-1-63483-166-6

Chapter 1

CHINA'S WIND AND SOLAR SECTORS: TRENDS IN DEPLOYMENT, MANUFACTURING, AND ENERGY POLICY*

Iacob Koch-Weser and Ethan Meick

EXECUTIVE SUMMARY

Renewable energy is gaining currency around the globe, but China and the United States are central to its development. They are the world's top-two countries in terms of energy consumption, net oil imports, and carbon emissions, as well as gross domestic product (GDP) and manufacturing. Their large territories harbor some of the best sites to generate renewable energy. If the United States and China cooperate—and compete—effectively, renewable energy can contribute to economic growth, energy security, and climate change mitigation. The past two years saw important developments, including a bilateral agreement to phase down hydrofluorocarbon emissions (June 2013)[1] and a joint announcement to reduce greenhouse gas emissions (November 2014).[2]

China is often touted for its rapid adoption of renewable energy technologies. Long-term industry plans and new legislation suggest Beijing will maintain this policy direction, in line with its overall expansion of energy

* This is an edited, reformatted and augmented version of a staff research report issued by the U.S.-China Economic and Security Review Commission, March 9, 2015.

production capacity. In the United States, by contrast, renewable energy is a divisive issue. There are disagreements about how, if at all, the government should support alternative energy sources when natural gas is abundant, emissions are declining, and energy demand is slowing.

This report assesses recent developments in China's wind and solar industries and the implications for the United States. It builds on the Commission's past work on U.S.-China energy issues, including the April 2014 hearing on bilateral clean energy cooperation.[1] The research also draws on Congressional testimonies, academic papers, industry and media reports, and statistical data. The report's main themes and findings are outlined below.

1) What have been the key successes and failures of the wind and solar industries in recent years, and what role has China played?

- Within a span of 15 years, the wind and solar industries have achieved significant increases in deployment, investment, manufacturing value-added, and goods exports. On all these metrics, China has become a leader. Its entry into the industry has helped reduce costs and pioneer modular supply chains that stimulate trade with U.S. companies. Several Chinese wind and solar manufacturers rank in the global top ten.
- However, China's rapid deployment of renewable technologies has also resulted in industrial overcapacities and a low rate of renewable power generation. Scores of solar PV equipment factories stand idle or underutilized. China's dumping of excess production onto world markets has given rise to protracted trade disputes with the United States and the European Union (EU). Although a rising share of wind and solar power in China is grid-connected, integrating that power into the electricity mix has been difficult. China is also missing its medium-term targets for offshore wind capacity.
- Because the United States accords greater weight to market forces, it has done a better job than China in ensuring renewable energy installed translates into power generated. Yet, its policy framework lacks comprehensive direction. U.S. states provide various levels of support, while federal programs such as the production tax credit have not been renewed by the U.S. Congress. Some skeptics in the United States argue taxpayers and utility companies should not subsidize "green" industries that cannot compete with conventional fuels on the open market.

2) What is China doing to address the shortcomings in its wind and solar policies?

- The Chinese government has attempted to “rebalance” its renewable energy sector. More priority is being given to domestic deployment of solar power, improving grid connectivity for wind farms, and reducing curtailment of renewable electricity by grid operators. “Mega wind farms” are being avoided in favor of smaller generation units near cities. Price supports for onshore wind have been scaled back as well for the first time in six years, and early efforts are being made to transfer the cost of renewable energy to electricity consumers, who in the past have benefited from low electricity bills. On the industry side, certain inefficient manufacturers have been allowed to fail. Successful companies, in turn, are being encouraged to establish a stronger presence in overseas markets and vertically integrate into research and development (R&D) and downstream utility sector services.
- Nevertheless, vested interests push back against such “rebalancing” policies in China. Absolute targets for renewable energy capacity distort the incentives of local officials. The integration of wind and solar into the grid is further impeded by China’s utility sector, which is characterized by monopolistic enterprises, price controls, and a preference for thermal power generation. Attempts by wind turbine suppliers to diversify their activities overseas have also had mixed success, as small emerging markets have proven easier to enter than the United States.

3) What is the impact of China’s wind and solar industries on global supply chains, the U.S. trade balance, and technological advances?

- To an extent, U.S.-China trade in renewable energy components is a “win-win” that fosters supply chain integration and mutual comparative advantage. Companies in both countries are innovating cutting-edge technologies such as permanent direct-drive turbines (China’s Xinjiang Goldwind Science and Technology Co.) and thin-film solar PV (the United States’ First Solar, Inc.).
- China is following the United States’ lead in adopting smart grid technology, thereby enhancing grid flexibility in ways that facilitate wind and solar power. China is also ahead of the United States in

building ultra-high voltage (UHV) transmission lines that reduce power loss over long distances. UHV lines help transmit wind and solar power generated in China's western regions to population centers further east.

- Despite bilateral cooperation, the Chinese government has pursued an aggressive industrial policy that aims to take market share in segments where the United States currently has an advantage. These segments include silicon and wafers in the solar PV sector, and blades and generating sets in the wind sector. China controls most production of rare earth elements (REE) integral to next-generation turbine technology, and beginning in 2009 imposed export restrictions. Production gains outside China have not completely resolved this problem. Following its defeat in a World Trade Organization (WTO) dispute involving the United States, Japan, and the EU, China stated in January 2015 it would lift these restrictions, though full compliance is uncertain at this stage.

4) How do wind and solar fit into China's energy security strategy and environmental policies?

- In terms of energy security, China faces greater challenges than the United States due to higher levels of GDP growth and energy intensity (energy per unit of GDP), a less balanced energy mix, and rising import dependence. The use of domestic coal reserves, as well as state-led efforts to increase overseas oil production and build energy stockpiles, is a partial solution at best.
- The Chinese government has started to prioritize demand-side energy efficiencies such as fuel standards, price liberalization, tax reform, and environmental performance requirements for local cadre officials. Set against these measures is the imperative to sustain economic growth.
- The expansion of gas, nuclear, and hydropower in China could eventually reduce state support for wind and solar. Yet, these alternative energy sources also come with tradeoffs. Acute water scarcities, social costs, and transboundary water disputes place constraints on hydropower generation. Shale gas extraction is challenged by difficult geology, environmental externalities, and uncompetitive bidding. Although China is introducing the latest in

Western nuclear technology, the rapid build-out of new reactors raises questions about safety.

- The United States and China have not made binding commitments at the multilateral level to reduce carbon emissions. However, they have jointly announced their commitment to emissions reductions. The United States is pushing its agenda via *The President's Climate Action Plan*, while China is pursuing energy and carbon intensity targets, in addition to establishing a domestic carbon trading platform. An added incentive for China to reduce emissions is in order to improve air quality, which has become a serious public health hazard.

5) What are the primary implications for the United States?

- Because wind and solar are emerging industries, the U.S. government is forced to strike a delicate balance, holding China accountable for unfair trade practices without disrupting the bilateral trade and cooperation that benefits the industry as a whole.
- Acquiring lower-cost clean energy equipment from China may be economical for U.S. importers in the short run, but could eventually prove costly if the equipment is of low quality.
- China's shift from labor- to capital-intensive production could bode well for some U.S. companies. Greater household wealth could make Chinese consumers more inclined to pay for wind and solar electricity and stimulate demand for high-quality U.S. imports. At the same time, China's upgrading into higher-margin segments of the wind and solar supply chain may intensify competition.
- If a binding, multilateral commitment to future carbon reductions is made at this December's United Nations Climate Change Conference in Paris, it could give renewed impetus to wind and solar deployment.

This report commences with an overview of the wind and solar industries' recent successes and failures. Section 2 evaluates China's measures to "rebalance" renewable energy deployment. Section 3 analyzes supply chains and technological developments, and how these are impacting U.S.-China trade in renewable energy equipment. The concluding sections look at renewable energy in the context of energy security and the environment, and consider implications for the United States.

SECTION 1: THE PERFORMANCE OF THE WIND AND SOLAR INDUSTRIES

Key Successes

The wind and solar industries have come a long way in a short period. According to a study by Pew Charitable Trusts, 735 gigawatts (GW) of clean energy technology were installed worldwide by year-end 2013, compared with virtually zero at the turn of the century. Wind accounted for the biggest share (307 GW), followed by small-scale hydropower (196 GW) and solar (144 GW).[3] China and the United States topped the list, together comprising 44.8 percent of the world's installed clean technology (see Table 1). The U.S. Department of Energy (DOE) calculates that deployed wind power capacity in the United States is now equivalent to the generation capacity of 60 large nuclear reactors.[4] In the solar sector, the United States and China's combined share of global installed capacity increased by 7.9 percentage points in 2012–2013. Germany, the solar market leader, saw its share decline.

At the time of this report's publication, Pew had not yet provided comprehensive cross-country data for wind and solar power capacity additions in 2014. New numbers from Bloomberg New Energy Finance (BNEF), however, affirm that China and the United States continue to lead the wind energy sector. The United States added 4.7 GW— a six-fold improvement over 2013—and China 20.7 GW, a gain of 23.7 percent year-on-year.[5]

The industry has also received substantial capital. The 2014 Pew study states annual clean energy investment worldwide—comprising wind, solar, and other sources—averaged $242 billion in 2011–2013, compared with a mean of $101 billion in 2005–2007.[6] BNEF reports global clean energy investment totaled $310 billion in 2014, the majority in wind and solar (see Table 2). As with installed generating capacity, China and the United States lead the pack, accounting for 45.6 percent of total clean energy investment last year.[7]

Table 1. Cumulative Clean Technology Installed, 2013
(Gigawatts; share %)

	Gigawatts				Share (%)				Year-on-year change (%)			
	Total	Wind	Solar	Other	Total	Wind	Solar	Other	Total	Wind	Solar	Other
WORLD	**735.0**	**307.0**	**144.0**	**284.0**	**100%**	**100%**	**100%**	**100%**	**13.4%**	**9.6%**	**38.5%**	**7.6%**
China	191.0	88.6	19.1	83.3	26.0%	28.9%	13.3%	29.3%	25.7%	19.7%	193.8%	16.5%
United States	138.2	59.4	12.4	66.4	18.8%	19.3%	8.6%	23.4%	3.9%	0.0%	53.1%	1.4%
Germany	77.4	34.0	35.5	7.9	10.5%	11.1%	24.7%	2.8%	9.0%	9.7%	10.9%	-1.2%
India	30.0	18.2	2.4	9.4	4.1%	5.9%	1.7%	3.3%	-1.6%	-9.0%	84.6%	2.2%
Rest of World	298.4	106.8	74.6	117.0	40.6%	34.8%	51.8%	41.2%	14.1%	11.7%	33.0%	6.6%

Source: Pew Charitable Trusts, 2013: Who's Winning the Clean Energy Race? (April 2014); Pew Charitable Trusts, 2012: Who's Winning the Clean Energy Race? (April 2013). [2]

Table 2. Clean Energy Investment, 2013–2014 (US$ billions)

	By Country					By Energy Source			
	2013		2014			2013		2014	
	$ bn	% share	$ bn	% share		$ bn	% share	$ bn	% share
China	$ 68	25.3%	$ 90	28.9%	Solar	$ 120	44.6%	$ 150	48.3%
US	$ 48	17.9%	$ 52	16.7%	Wind	$ 90	33.4%	$ 100	32.1%
Other	$ 152	56.8%	$ 169	54.4%	Other	$ 59	21.9%	$ 61	19.6%
Total	$ 268	100.0%	$ 310	100.0%	Total	$ 268	100.0%	$ 310	100.0%

Note: Rounding errors. Shares may not add up to 100 percent. *Source*: Bloomberg New Energy Finance.

Wind and solar energy have also become more price-competitive.

- BNEF estimates turbine prices fell by 29 percent and solar PV prices by 80 percent between 2008 and 2013.[8]
- According to the DOE, leading up to 2009, the national average levelized cost of electricity (LCOE) [3] of new wind power purchasing agreements in the United States fell to a record low of $25 per megawatt-hour (MWh). In the solar sector, rooftop solar panels now cost 1 percent of what they did 35 years ago, and the cost for a PV module declined from $3.40/watt in 2008 to $0.79/watt in 2014.[9] The U.S. Energy Information Administration (EIA) estimates that for plants entering service in 2019, the LCOE per MWh of onshore wind will be $80.30, cheaper than conventional coal ($95.60).[10]
- China has experienced a substantial drop in the per-unit cost of turbines in wind farm concession bidding. For example, at the auction of 10 GW-scale wind power bases in Hebei Province and Xinjiang Autonomous Region in 2012, one of the successful bids for 1.5 MW-scale turbines quoted a price of $554 per MW, a decrease of $339 over the price submitted in a 2008 auction.[11]

Technological breakthroughs are a principal driver of cost reductions. Larger and taller wind turbines are now capable of generating steadier output. A 2014 DOE report states that, compared with 1998–1999, the average electricity generating capacity of a single turbine in the United States has increased by 162 percent.[12] In China, average turbine size installed jumped from 850 kilowatts (kW) in 2009 to 1,393 kW in 2012.[13]

Important as well are improvements in manufacturing processes and business models. The DOE calculates that doubling solar PV manufacturing capacity translates into a 20 percent decline in PV prices.[14] Production occurs across national borders and along complex value chains, introducing scale efficiencies and opportunities for cooperation. Solar PV modules exported from China, for example, may be produced using intellectual property from Germany, capital equipment from the United States, and cells converted in Taiwan.

By playing a central role in these improvements and innovations, China has been able to "leapfrog" into a nascent industry not yet dominated by Western companies. At the 2009 World Economic Forum, China's then premier Wen Jiabao announced that "China will accelerate the development of a low-carbon economy and green economy so as to gain an advantageous position in international industrial competition."[15] China's insertion into the global solar PV industry really took hold in the mid-2000s, when there was a spike in European demand. China came to the industry much later than European and U.S. solar firms [4] but was able to undercut the market with cheaper panels at acceptable levels of quality, aided by government subsidies and a low cost of labor and other factor inputs. In 2013, six out of the world's top-ten solar PV manufacturers were located in China, compared with only one U.S. firm, Arizona-based First Solar. [5]

Through a combination of market competitiveness and strong government backing, China has made similar strides in the wind turbine industry. Eight of the world's top-15 wind turbine makers in 2013 were from China. U.S. and European incumbents, though, have the prevailing market share outside Asia; among them is the U.S. multinational conglomerate General Electric (GE). A fundamental difference between the wind and solar industries is that turbine components are less tradable, due to their bulk and the need to tailor turbines to local wind conditions.

The wind and solar industries can generate jobs at laboratories, factories, and power generation sites, as well as in corollary sectors such as insurance and finance. Data on sales of low-carbon goods and services, compiled by the United Kingdom's (UK) Department for Business Innovation and Skills, ranked the United States first and China second in the world in 2012 (see Table 3). [6] A 2014 report by the Renewable Energy Policy Network for the 21st Century (REN21) counts 6.49 million jobs in the renewable energy sector worldwide, with half related to wind and solar. China is the chief job creator (41 percent share). The United States ranks third (10 percent share) behind Brazil, a leader in hydropower and biofuels.[16]

Table 3. Low-Carbon Environmental Goods and Services by Country, 2008–2012 (UK Government Estimate) (US$ billions)

		Total Sales (US$ billions)				Share (%)			
		2008/09	2009/10	2010/11	2011/12	2008/09	2009/10	2010/11	2011/12
	Total	$ 1,959	$ 2,067	$ 2,131	$ 2,128	100.0%	100.0%	100.0%	100.0%
1	US	$ 395	$ 407	$ 415	$ 409	20.1%	19.7%	19.5%	19.2%
2	China	$ 261	$ 276	$ 280	$ 275	13.3%	13.3%	13.1%	12.9%
3	Japan	$ 123	$ 128	$ 132	$ 132	6.3%	6.2%	6.2%	6.2%
4	India	$ 121	$ 129	$ 132	$ 130	6.2%	6.2%	6.2%	6.1%
5	Germany	$ 82	$ 88	$ 90	$ 90	4.2%	4.2%	4.2%	4.2%
	Top-10	$ 1,269	$ 1,337	$ 1,369	$ 1,356	64.8%	64.7%	64.2%	63.7%
	Rest of World	$ 690	$ 729	$ 763	$ 773	35.2%	35.3%	35.8%	36.3%

Source: United Kingdom Department for Business Innovation and Skills, Low Carbon Environmental Goods and Services: 2011 to 2012 (July 2013). https://www.gov.uk/government/publications/low-carbon-and-environmental-goods-and-services-2011-to-2012.

Key Challenges

Deployment Issues

While the wind and solar sectors have made significant progress, serious challenges persist. To begin with, installed capacity is not synonymous with power generation, especially in China. Wind and solar together accounted for less than 3 percent of China's electricity generation in 2013 (solar is probably still less than 0.5 percent). Thermal power—predominately coal—comprises the bulk of new power generation (see Figures 1 and 2).[17] This lack of renewable power generation is striking considering China's dominance as a wind and solar equipment manufacturer.

The expansion of coal-fired power has diminished the relative impact of wind and solar, but so have problems inherent to the renewable energy sector. As China's installed wind capacity has skyrocketed, many wind generators remain unconnected to the grid. According to the China Renewable Energy Industries Association (CREIA), between 2010 and 2012, up to one-third of installed wind capacity in China was not grid-connected.[18] Yu Guiyong, an expert at the China Wind Energy Association, claims China in 2013 connected only 48 percent of newly added wind capacity.[19] Wind farms have proliferated in China's western regions, far from the eastern seaboard where the bulk of electricity is consumed. [7] Utility companies have been slow to make the necessary investments in dedicated wind farm transmission lines. Similar problems plague China's nascent solar power sector.[20]

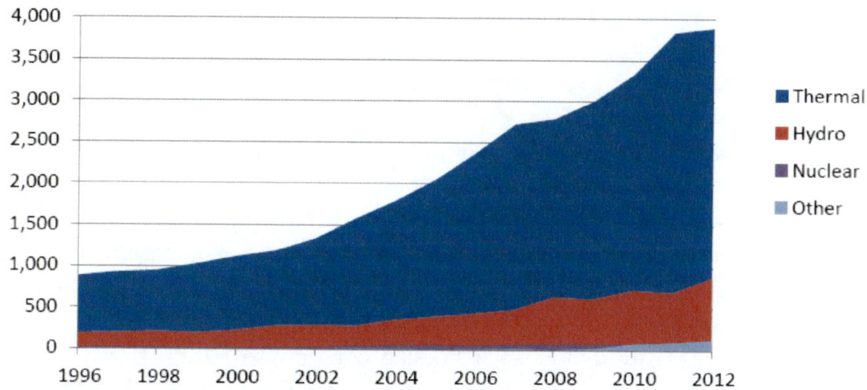

Note: "Other" is mainly wind power. "Thermal" is mainly coal; the natural gas and petroleum share of thermal power generation is very low.
Source: China National Bureau of Statistics, via CEIC.

Figure 1. China's Power Generation by Source, 1996–2012. (Gigawatt-hour [GWh] thousands).

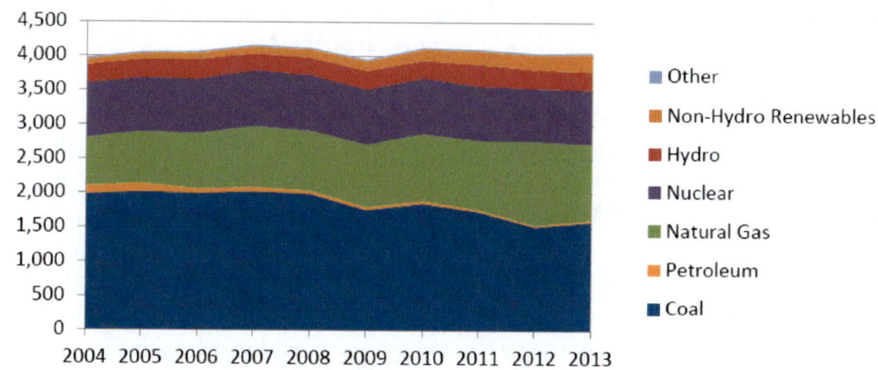

Source: U.S. Energy Information Administration.

Figure 2. U.S. Power Generation by Source, 2004–2013. (GWh thousands).

The latest data, published by China's National Energy Administration (NEA), implies some improvement in wind power grid connectivity. The agency claims China connected 18.7 GW in 2014, for a cumulative total of 96

GW of grid-connected wind power. If cumulative installed capacity last year reached 109 GW (as BNEF data indicates), then China's connectivity ratio is now a respectable 88.1 percent.[21]

However, the connectivity problem should not be viewed in isolation. For one, wind farms that do connect to the grid are often underutilized. In 2013, the ten Chinese provinces with the most installed wind capacity had a wind turbine capacity factor [8] ranging from 18 percent (Jilin province) to 29 percent (Xinjiang Autonomous Region). The U.S. average was 33 percent.[22] Datang Renewables—a unit of one of China's state-owned power generation companies—said utilization rates at its wind farms throughout China dropped dramatically in the first half of 2014 owing to a fall in wind speeds.[23]

Due to poor quality, turbines installed in China also malfunction too frequently. Other countries certainly face this problem, but the consequences are particularly severe for China's mega wind farms. For instance, the Northwest Power Grid in the Gansu Jiuquan wind power base suffered a short circuit fault in February 2011 that led to a massive output loss of 840 MW from 598 turbines.[24]

Such incidents threaten the safe operation of the power grid, and reinforce the tendency of grid operators to avoid wind power where possible. Indeed, curtailment—whereby a grid operator refuses to use electricity from a power generator—is a problem as well. As much as 11 percent of China's wind power was curtailed in 2013, compared to rates of 1 to 4 percent in the United States.[25] China's NEA claims curtailment rates fell to 7.5 percent through the first nine months of 2014, a mark of steady improvement.[26] Nonetheless, the problem has not been resolved, and it is too early to tell whether the recent drop is owing to cyclical trends or structural improvement.

Due to wind power's intermittency, as well as its tendency to maximize output during non-peak periods of electricity demand, utility-scale deployment may force grid operators to expand baseload capacity [9] to account for shortfalls.[27] Such measures may be acceptable if the energy source compensating for wind is clean and adaptable; in Denmark, for example, grid operators import hydropower from neighboring countries.[28] Yet in countries such as China and the United States, a ready source of baseload power is coal. In addition to being polluting, coal-fired generators take time to shut off and restart, making them less compatible with wind and solar energy.

The United States probably still generates more power from wind than China, even though it has less installed capacity (see Figure 3). [10] Non-hydro renewable power comprised 6.2 percent of U.S. power generation in 2013, a threefold increase over 2004, primarily thanks to wind power.[29] John

Romankiewicz, a clean energy analyst affiliated with University of California-Berkeley, predicts China will not surpass the United States in total wind power generation before 2020.[30]

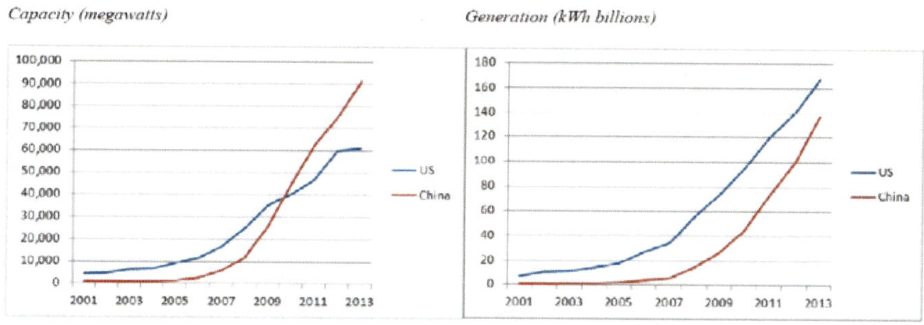

Sources: U.S. DOE Energy Efficiency and Renewable Energy Office; Global Wind Energy Council; Greenridge Global; and the U.S. Energy Information Administration.

Figure 3. Wind Power in the United States and China: Capacity vs. Generation, 2000–2013. (Megawatts; kWh billions).

The dilemma for the United States is the "boom and bust" cycle of its wind and solar industries. Tax credit subsidies for manufacturing, investment, and power generation require periodic approval from the U.S. Congress. When approval is uncertain, investors become reluctant to commence new wind and solar projects that hinge on tax credits to be profitable.[31] According to BNEF, the United States in 2012 ranked a mere tenth among the Group of 20 major economies (G20) in clean energy invested per dollar of GDP (renewable investment intensity), as investment plummeted by 37 percent year-on-year in 2011–2012 due to the imminent expiration of the production tax credit.[32] Just three years earlier, the American Recovery and Reinvestment Act of 2009 constituted the biggest ever public sector investment in the U.S. clean energy sector.[33]

Policy uncertainty in the United States illustrates an underlying problem—dependence on government support to make renewable energy viable. Costs have certainly declined, especially for onshore wind farms. But governments still provide strong backing, either directly in the form of fiscal spending and tax incentives, or indirectly through regulations such as consumer surcharges, loan guarantees, and renewable portfolio standards. [11] A central disagreement concerns price subsidies for wind and solar deployment. Supporters argue such subsidies are small relative to the private investment they stimulate, and in fact help reduce technology costs over time.[34] They may also create a level playing field for nascent technologies, as mature technologies benefit from existing infrastructure and have a "sticky cost advantage."[35] Skeptics contend wind and solar technologies should be left to compete on the open market. Professor Robert J. Michaels, an economist at California State University-Fullerton, told Congress that wind power is not worth funding with tax credits because it is too expensive, decreases the efficiency of grids, and distorts the investment decisions of utility companies.[36]

Offshore wind is an emerging area singled out for government support. A record $19.4 billion was invested in this sector worldwide in 2014.[37] The EU, in particular the United Kingdom, has pioneered offshore wind development and leads the world in installed capacity.[38] The European Wind Energy Association forecasts offshore wind power to account for 4 percent of the EU's total electricity demand in 2020, and 14 percent in 2030.[39] In the United States, a 2011 report by the DOE and the Department of the Interior outlines plans to deploy 54 GW of offshore wind by 2030, about four times what the EU currently has installed and under development.[40] China is equally ambitious. The government's 12th Five-Year Plan (2011–2015) targets 5 GW of offshore wind capacity by 2015 and 30 GW by 2020. China built the world's first utility-scale offshore development outside of Europe in July 2010, giving rise to other pilot projects along the coastlines of Shanghai municipality and Jiangsu province.[41]

The road to offshore wind has been rocky, however. A July 2014 Bloomberg report finds China is three years behind its plan to install 5 GW of offshore capacity by 2015; less than 10 percent had been installed at that point.[42] Companies in Europe have also scrapped plans for more than 5,700 MW since November 2014.[43] In the United States, offshore wind has barely gotten off the ground. The first U.S. grid-connected offshore wind pilot project, located in Maine, took until 2013 to complete.[44] Cape Wind Associates LLP has struggled for a decade to build the country's first offshore wind farm in Nantucket Sound, off the coast of Massachusetts. As of January 2015, the project was in the "financing and final commercial contracting stage."[45]

Table 4. Estimated Capital Costs for Onshore and Offshore Wind Power Systems, 2011

	Onshore	Offshore
Capital investment costs ($/kW)	1,700-2,450	3,300-5,000
Wind turbine cost share (%)	65 to 84	30 to 50
Grid connection cost share (%)	9 to 14	15 to 30
Construction cost share (%)	4 to 16	15 to 25
Other capital costs (%)	4 to 10	8 to 30

Source: International Renewable Energy Agency, via Ross McCracken, "Wind Giants," *Platts Energy Economist*, January 1, 2014, via Factiva.

Aside from issues concerning jurisdiction over maritime areas, offshore wind suffers from the cost burden of constructing and connecting turbines in the sea (see Table 4 above).

The EIA estimates new offshore wind farms coming online in 2019 will cost $204.10 per MWh—over three times the price of conventional gas-fired power.[46] As such, offshore wind will likely require government support for years to come.

Manufacturing Issues

The renewable energy equipment sector presents a separate set of challenges. Although the recent drop in solar PV prices has raised the appeal of solar power for consumers and grid operators, price declines are a byproduct of excess production in China. In early 2013, over half of all solar companies in Jiangsu province, where two-thirds of China's solar PV panels are produced, were forced to halt production due to the sharp fall in panel prices and reduced demand from Europe and the United States.[47]

China's overcapacities in the solar PV sector have wasted resources, including polysilicon inputs originating from the United States. They have also crowded out U.S. competitors; the U.S. share of global PV manufacturing fell from 43 percent in 1995 to 27 percent in 2000 and just 7 percent in 2010.[48] U.S. Census data shows U.S. imports of Chinese solar cells and modules jumped in 2011, accompanied by a sharp drop in the unit price.[49] Testimony from U.S. solar PV makers, provided to the U.S.-China Economic and Security Review Commission in July 2010 and to the International Trade Commission in October 2012, points to substantial job loss—particularly in Ohio— due to cheap equipment imports from China.[50]

Critics claim the Chinese government is unfairly supporting its clean energy equipment producers. Rob Atkinson, president of the Information

Technology and Innovation Foundation (ITIF), stated in 2011 that China is practicing “a very aggressive form of innovation mercantilism” that harms the clean energy sector and U.S. interests.[51] [12] Market-distorting policies arguably include:

- *Extensive subsidies.* In the course of investigations to evaluate whether the United States should impose anti-dumping and countervailing duty (AD/CVD) measures against Chinese solar PV producers, the U.S. Department of Commerce (Commerce) found that the Chinese government offers a range of subsidies, in the form of grant programs; government provisions of goods and services for less than adequate remuneration (primarily polysilicon, aluminum, and electricity); income and other tax exemption and reduction programs; and export subsidies and guarantees.[52] A 2010 National Foreign Trade Council study, authored by the law firm Dewey & Leboeuf LLP, takes note of the NDRC’s *Guidance Catalogue on Renewable Energy Industrial Development* (issued November 2005), which offers preferential tax treatment or designated funding to 88 types of renewable energy projects.[53]
- *Induced demand for domestic producers.* The National Foreign Trade Council report states that China’s bidding procedures for large wind farms favor China’s state-owned power generation companies, who in turn procure domestically produced equipment.[54] News reports also suggest the Chinese government creates demand for domestic equipment producers via diplomatic initiatives. For example, during the 2010 state visit of then Premier Wen Jiabao to Pakistan, a project to generate 2.3 GW of wind and solar power was the most important agreement signed by the two countries.[55] In the spring of 2013, Premier Li Keqiang actively lobbied in European capitals against the European Commission’s efforts to levy antidumping duties on Chinese solar panels. [13]
- *State-backed financing.* Certain Chinese companies have had generous access to capital from state-owned banks and funds. Examples of financing deals are China Development Bank’s $5.5 billion loan to turbine maker Goldwind for the 2011–2015 period,[56] a $4.8 billion loan to China Guangdong Nuclear Power Holding Co. to develop wind power projects at home and abroad,[57] and individual loans to developers of overseas projects in return for procuring China-brand turbines.[58]

- *Intellectual property appropriation.* Chinese equipment producers have in some instances acquired intellectual property through outright theft, and as is more commonly the case, leveraged market access to compel technology transfer from foreign wind turbine companies selling into the China market.[59]

To address China's unfair trade practices, the Obama Administration has taken concerted action through trade remedies: [14]

- In October 2010, Commerce notified the WTO of almost 200 Chinese clean energy subsidies in violation of trade rules.[60]
- In November 2012, Commerce imposed anti-dumping duties on all imports of Chinese PV cells and modules.[61] The duties did not, however, apply to solar panels assembled in China with components from third countries. To take advantage of this loophole, many Chinese firms moved production of solar cells to Taiwan while maintaining panel assembly in China. [62] SolarWorld Americas, the original petitioner, complained this loophole weakened the effect of the original duties.[63]
- In December 2014, Commerce attempted to close the earlier loophole by introducing a new set of AD/CVD measures that capture China-origin solar panels made with solar cells from a third country.[64] Weeks later, Commerce confirmed it would maintain the 2012 AD duties as well, though it reduced the duty levels based on updated evidence of lower dumping margins by Chinese panel makers.[65]

Opponents of such trade remedies argue they are counterproductive. Michael Levi, an energy expert at the Council on Foreign Relations, finds barriers to clean energy imports may slow the flow of clean energy technology across borders, stifle innovation, starve capital-hungry U.S. firms of investment, and deprive U.S. consumers of access to cheaper sources of pollution-free power.[66]

In July 2013, China applied antidumping duties on U.S. polysilicon exports, a move interpreted by some as retaliation for the U.S. AD duties on solar panels.[67] Others assert that the U.S. government also provides support to its domestic producers, and that China has already rolled back certain support measures.[68]

SECTION 2: THE "REBALANCING" OF CHINA'S RENEWABLE ENERGY DEPLOYMENT

Changes in Deployment Strategy

China's wind and solar markets have developed in distinct phases. In the initial phase, dating back to the 1990s, domestic renewable energy deployment was marginal, and the majority of equipment was purchased from abroad. In the ensuing phase, beginning in the early to mid-2000s, domestic equipment manufacturing boomed and imports declined. This gave rise to a pattern in which China-made wind turbines were deployed domestically, while solar PV modules were exported. Consequently, China became the global leader in installed wind capacity as well as solar PV exports.

The government enacted a variety of measures to encourage wind deployment.

- *Concession bidding.* The NDRC, China's chief industrial policymaking agency, instituted a wind bidding process in 2003 for all wind farms greater than 50 MW. Winning bids were based on the average price of the turbine and at least 50 percent Chinese-manufactured content (a policy that remained in place until 2010). Most of the bidding was for mega wind farms situated in China's wind-abundant northwestern region.[69]
- *Mandatory grid access.* The 2006 Renewable Energy Law set renewable portfolio standards for utility companies, obligating them to purchase all of the wind energy produced and to connect wind farms to the grid.[70]
- *Feed-in-tariff (FiT).* The government introduced a nationwide FiT for wind power in 2009 with four regional benchmark prices. It used the earlier system of wind farm bidding prices as a point of reference.[71]
- *Capacity targets.* The NDRC's Medium- and Long-term Development Plan for Renewable Energy (2007– 2020) set a target of 30 GW grid-connected wind capacity by 2020.[72] By 2010, China had eclipsed the 2020 target ahead of schedule.[73] (The current target for 2020 is 200 GW.[74])

In contrast, China's solar energy policies initially were not geared toward domestic deployment. Rather, the decision of European governments to adopt

FiT policies for solar energy created new demand for solar panels. Chinese manufacturers, aided by export-oriented economic policies, seized the opportunity, becoming the dominant global suppliers of solar PV within a five-year span beginning in the mid-2000s. The years 2007–2008 were crucial, as solar manufacturers constructed industrial-scale facilities to produce PV wafers, cells, and modules.[75] Even as individual companies like Trina Solar became global leaders, hundreds of smaller enterprises flooded the market.

Around 2009–2010, however, imbalances started to appear. Excess production of renewable energy equipment, combined with reduced solar subsidies and aggressive trade remedies by trade partners, disincentivized solar PV exports. Domestically, many wind farms were underutilized.

This situation prompted the Chinese government to pursue a "rebalancing" policy, characterized foremost by a shift in domestic deployment toward the solar sector. Although China's installed wind capacity remains many multiples higher than solar, that ratio has declined dramatically since 2010 (see Figure 4). In 2012–2013, solar capacity installed grew by 193.8 percent year-on-year, versus 19.7 percent growth for wind capacity. According to the estimated cumulative figures provided by REN21, China by 2014 ranked second in installed solar capacity behind only Germany.[76]

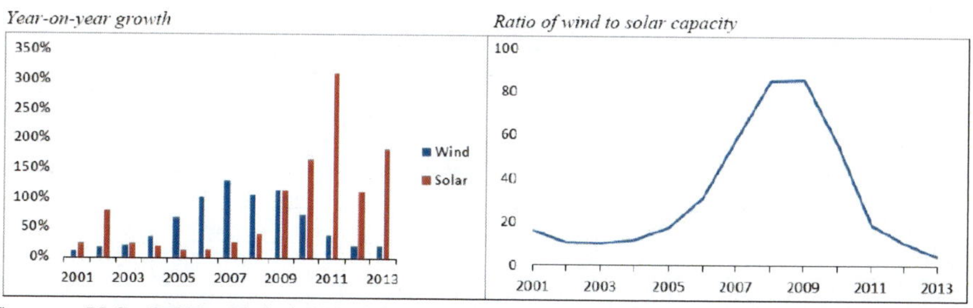

Source: U.S. DOE; Global Wind Energy Council; China State Electricity Regulatory Commission; China Electricity Council; International Energy Agency; and REN 21.

Figure 4. Installed Capacity of Wind and Solar Energy in China, 2001–2013. (Ratios and growth rates based on MW installed).

Following the lead of the wind sector, the government introduced a series of measures to promote solar deployment. The Solar Roofs program, unveiled in March 2009 under the direction of China's Ministry of Finance (MOF) and Ministry of Housing and Urban-Rural Development, provides subsidies for roof-mounted PV systems. The program accords priority to grid-connected solar projects in order to avoid the grid connectivity issues affecting remote wind farms. Also in 2009, the government created the Golden Sun program, which subsidizes solar PV manufacturing and installation, [15] along with a concession bidding program for utility-scale solar projects.[77]

China proceeded to introduce a solar FiT in 2011. Its point of reference was the 2009 provincial FiT in Jiangsu province, where most solar manufacturers are based. In conjunction with the solar FiT, the country's main grid operator, State Grid Corp. of China, expanded and upgraded its electricity grids to better connect solar power.[78] State Grid also announced a plan in October 2012 to allow distributed solar power generators [16] smaller than 6 MW to connect to its power lines.[79]

In parallel, the government has revised renewable energy deployment targets for solar power (see Table 5). In its *Guidance on Promoting the Healthy Development of the Solar Industry*, issued in July 2013, the State Council raised the solar capacity target to 35 GW by 2015, up from the 21 GW target set in the NEA's *2011–2015 Solar Development Plan*, issued just a year earlier.[80]

Table 5. China's Official Deployment Targets for Renewable Energy (Gigawatts, share %)

	Cumulative Capacity					Annual Additions			
	2012	2013	2014	2015	2020	2013	2014	2015	5-yr aver. ('15-'20)
	Gigawatts								
Hydro	212	233	261	289	430	21	28	28	28
Wind	48	66	83	100	200	18	17	17	20
Solar	3	13	24	35	50	10	11	11	3
Biomass	7	7	10	13	30	0	3	3	3
Total	270	319	378	437	710	49	59	59	55
	Share (%)								
Hydro	78.5%	73.0%	69.0%	66.1%	60.6%	42.9%	47.5%	47.5%	51.6%
Wind	17.8%	20.7%	22.0%	22.9%	28.2%	36.7%	28.8%	28.8%	36.6%
Solar	1.1%	4.1%	6.3%	8.0%	7.0%	20.4%	18.6%	18.6%	5.5%
Biomass	2.6%	2.2%	2.6%	3.0%	4.2%	0.0%	5.1%	5.1%	6.2%

Note: Due to frequent policy revisions, current targets may vary slightly from those listed here. Hydropower capacity goal includes 50 GW of pumped hydro storage.

Source: Adapted from World Resources Institute, "Why Is China Taking Action on Clean Energy and Climate Change?" ChinaFAQs Issue Brief (May 2013), p. 4.

Policy adjustments to wind deployment have been less drastic. In 2014, China installed nearly five times as much new onshore wind capacity as the United States, exceeding BNEF's earlier projections.[81] Still, there have been novel approaches. To tackle grid connectivity issues, the government started construction in 2009 on a 10 GW-level wind power station—nicknamed "Three Gorges in the Air"—with guaranteed access to transmission infrastructure.[82] As with solar, the NEA began in 2011 to endorse the dispersed development of smaller wind farms near urban centers, where the disadvantage of low wind speed is to be offset by savings on power transmission costs.[83]

Generous price supports are a key inducement to build wind farms. The NDRC proposed in September 2014 to cut per kilowatt-hour tariffs for onshore wind power for the first time since the national FiT was introduced in 2009.[84] The proposed cuts were deeper than the industry expected. Zhou Shiyi, an analyst at BNEF, predicted idled wind farm projects and a potential decline in new wind installations, as "developers building projects originally planned for 2016 or 2017 then won't have enough capital."[85] True to its word, the NDRC enacted tariff cuts in January 2015, to be applied to all wind projects approved after 1 January 2015 and to projects approved before that date but commissioned after 1 January 2016.[86] Closer analysis, however, reveals the cuts were not as deep as initially proposed, suggesting the government may have buckled under industry pressure (see Table 6).

Table 6. China's Onshore Wind Tariff Reductions: Proposed vs. Enacted Tariff Cuts (RMB)

		Cuts proposed Sep14		Cuts enacted Jan15	
	Original tariff	New tariff	Change	New tariff	Change
Region 1	0.51	0.47	0.04	0.49	0.02
Region 2	0.54	0.50	0.04	0.52	0.02
Region 3	0.58	0.54	0.04	0.56	0.02
Region 4	0.61	0.59	0.02	0.61	0.00

Source: Eric Ng, "Wind Power and Equipment Maker Shares Slide 6pc by Midday," South China Morning Post, September 16, 2014. http://www.scmp.com /business/commodities/article/1593078/windpower-and-equipment-maker-shares-slide-6pc-midday; Jianxiang Yang, "China Confirms Cut to Onshore Wind Tariff," Wind Power Monthly, January 8, 2015. http://www.wind power monthly.com/article/1328437/china-confirms-cut-onshore-tariff.

As noted in Section 1, the most recent data for 2014 suggests improved connectivity and reduced curtailment.[87] Zhu Ming, deputy director of the National Energy Administration (NEA), pronounced last October that the government would "introduce new measures to eliminate curtailment in 2015." He did not specify what precisely those measures would look like.[88]

Structural Problems in China's Utility Sector

The utility sector presents a stubborn hurdle to "rebalancing" China's renewable energy deployment. Since abandoning a command economy in the 1980s, the Chinese government has struggled to create a functioning market that provides adequate and cheap supplies of electricity to factories and cities. A rise in power outages at the turn of the century prompted a radical overhaul of the utility sector in 2002–2004. It led to the creation of an independent electricity regulator, the State Electricity Regulatory Commission (SERC), and divided up the state's power monopoly into five state-owned power generation companies (the "Big Five") and two transmission and distribution companies—State Grid and the smaller China Southern Power Grid. The government also introduced a new pricing formula: power tariffs were to be periodically adjusted in accordance with the six-month average floating price of coal; split into wholesale and retail prices; assigned tariff levels by region; and charged separate taxes by province.[89]

These reforms marked a modest departure from the command economy but kept in place elements of the old system. The regional division of power generation and transmission assets among different state-owned enterprises (SOEs), coupled with local price and tax regimes, has hampered the integration of electricity markets.[90] At the same time, SERC has deferred to the government's planning body, the NDRC, which sets prices, devises long-term energy targets, and approves the major power generation facilities.[91]

In the years following the reforms, electricity price controls also remained largely intact. Coal prices rose exponentially in the 2006–2012 period, due to the upstream price reforms and the rapid growth in Chinese energy consumption. This should have resulted in a commensurate increase in the wholesale and retail electricity price, but NDRC officials balked at making the adjustments. Faced with periodic price squeezes, power generators incurred substantial losses, and many generation sites ran at below capacity due to high fuel costs.[92] In parallel, Chinese energy consumers were shielded from market prices. Cross-country data for 2012, aggregated by the World Energy Council, puts the cost of electricity per kilowatt-hour at five cents in China—versus 12 cents in the United States and 28 cents in Japan.[93]

Over the past three years, Beijing has stepped up its efforts to reform electricity prices. The NDRC introduced a three-tier residential electric price in July 2012, designed to make customers who use more power pay higher rates per kilowatt-hour than those who use less.[94] According to Ryan Rutkowski, an economist at the Peterson Institute for International Economics, the price adjustment has helped utility companies earn a higher return on assets, which could incentivize them to turn away from cheap coal-fired power. However, it is not clear whether this reform trajectory can be maintained:

> In 2008 to 2012 [prior to the price reform], returns [on assets of utility companies] were suppressed by the state as part of a broader effort to stimulate the economy following the financial crisis. If coal prices rise once again, the state still has complete control over end user pricing, and thus can limit pass-through of rising on-grid tariffs. The 5 percent threshold for annual adjustment to on-grid tariffs also leaves plenty of room to squeeze thermal power producers once again. On the other hand, if market price reforms to electricity pricing continue, this will likely lead to higher prices. Higher prices will be an additional challenge to a manufacturing sector already struggling with rising labor, capital, and exchange rate costs.[95]

Institutional faults in the utility sector have negative implications for wind and solar deployment.

- *Prioritization of thermal power to expand generation capacity.* Overuse of cheap electricity on the demand side places enormous pressure on China's power generators and grid operators to ensure adequate supplies. The simplest way to do so and earn a profit is by expanding power capacity using domestic coal supplies. In this context, renewable energy mandates can be a burden. A common ploy for the Big Five power companies to comply with government directives has been to "cross-subsidize" wind and solar projects using the profits earned from conventional power.[96]
- *Dominance of SOEs.* Curtailing wind power is technically illegal, since utilities should be buying and distributing the wind power generated. Due to their market dominance and state ownership, however, China's grid operators—mostly subsidiaries of State Grid—

can do so with near impunity. Moreover, according to CREIA, a group of 700 SOEs across China accounted for 90 percent of all wind power projects and 79 percent of all grid-connected wind capacity in 2011. Within this group, the Big Five power generation companies comprised 71 percent of the total share. The Big Five are particularly prominent in wind farm projects above 50 MW, which the central government reserves the right to approve.[97]

- *Ineffective planning at the local level.* Local governments are permitted to build smaller wind farms below 50 MW without central government approval. This has given foreign turbine suppliers a point of entry into the Chinese market. But since such projects are implemented by local governments, the quality of local wind resources and transmission infrastructure is inconsistent. Some local governments have also forced developers to source equipment locally in exchange for wind farm permitting.[98] In fact, part of the motivation behind deploying renewable energy in China's northwest is to promote regional economic development, which has an underlying social stability dimension. Low grid connectivity and greater efficiency may therefore be tolerated for longer than if the objective were merely energy production.
- *Poor budgeting decisions.* The funds China has set aside via its National Renewable Energy Fund (under the 12th Five-Year Plan) may be insufficient to cover the future costs of the solar FiT. Forecasts indicate the targets can only be met if the fund's resources are exceeded by $13 billion every year through the end of the 12th Five-Year Plan in 2015. Li Shuo, a climate and energy campaigner at Greenpeace East Asia, predicted in 2013 that China might have to double the renewable energy surcharge to plug the gap between fiscal capacity and planned subsidies.[99] Currently, electricity prices are being liberalized in the context of low energy prices. A key challenge will be to raise surcharges once energy prices rebound.

A legacy of China's command approach to energy markets is the insistence on absolute targets. Central planners devise capacity additions and long-term targets for each energy source in multiyear comprehensive and sectoral plans. Such targets may create stable expectations for future capacity growth, without the messy variation in state-level policies one finds in the United States; [17] and yet, policymakers in advanced economies tend to dislike them because they distort incentives. Local officials may try to meet capacity targets through uncoordinated expansion or number fudging, neither of which contribute to substantive improvements.

Setting absolute targets for the wind sector appears especially ill-advised because the Chinese government lacks comprehensive knowledge about the country's wind resources. The United Nations Environmental Program's Solar and Wind Energy Resource Assessment (SWERA) has mapped wind resources in several developing countries, but only partial estimates are available for China's vast territory.[100] As engineering professor David Keith of Harvard University points out, there is disagreement even among international experts about how much wind capacity the world can install, given constraints on land use and other factors. [18] Professor Joanna Lewis of Georgetown University states:

> Many of China's policies that governed wind power development over the past decade were informed by the government's understanding of China's total wind resource potential and where the best wind resource sites were located. Official Chinese wind resource measurements have changed rather dramatically over time, however, and as new studies increasingly elucidate better data on these three topics, it is becoming clear that some of the early policies may have been based on assumptions that were incorrect. For example, it is likely that China's onshore resources were dramatically understated in early assessments, while China's offshore resources were substantially overstated.[101]

China's planners are unlikely to abandon absolute targets in the near term, but there are signs of a changing mindset. For one, the targets are becoming less rigid. The Energy Research Institute, a think tank under the NDRC, forecast last November that China's solar PV installations would reach 10 GW in 2014. Months earlier, NEA head Wu Xinxiong had given a target of 13 GW.[102] The NDRC's statement carries significant implications, since it suggests China might not meet the 2020 solar target of 50 GW.

More importantly, the government is placing a stronger emphasis on percentage and intensity-based targets. The 11th Five-Year Plan (2006-2010) for the first time formulated a target to lower energy intensity, supplemented in the 12th Five-Year Plan by specific pledges to cut carbon intensity and elevate the share of non-fossil fuels in primary energy consumption. At the U.S.-China summit last November, President Xi Jinping announced China's goal to bring the non-fossil fuel share to 20 percent by 2030, nearly twice the 2015 target.[103]

SECTION 3: TRENDS IN WIND AND SOLAR MANUFACTURING

Outbound Investment and Industry Consolidation

In conjunction with changes to deployment policy, China has modified its industrial policy targeting wind and solar manufacturers. One facet is outbound investment. Under the "Go Out" strategy, launched during the 10th Five-Year Plan (2001–2005), Chinese firms have been encouraged to seek markets, resources, and technology abroad. Corporate wealth in China is increasing rapidly, as are the People's Bank of China's holdings of foreign exchange reserves. Loose credit policies in the wake of the global financial crisis added further impetus to outbound flows. A report by the World Resources Institute (WRI), covering the years 2002–2011, indicates that these outbound investment policies are also influencing the wind and solar industry.

- Chinese solar and wind companies made at least 124 investments across 33 countries, with the United States as the prime recipient. The number of outbound investment projects by Chinese wind and solar companies doubled in 2009–2010. The 54 investments for which detailed data is available amounted to nearly $40 billion.
- Eighty-one projects were in the solar sector, and 41 were in the wind sector. Of the wind projects, 63 percent went toward wind farm development. Since turbines are bulkier than solar panels, turbine makers have an incentive to manufacture at or near overseas markets and to source some parts locally.
- Solar sector investments were spread among a dozen Chinese companies, nearly all of them producers of solar panels. In the wind sector, on the other hand, one company, Goldwind, accounted for 12 out of 41 projects. Utility sector SOEs are also leveraging their market position to diversify into the wind sector, including overseas investments. Of the 41 overseas wind investments recorded by WRI, 13 were made by power generation companies. Guodian United Power, for example, has acquired its own turbine manufacturing business.[104]

In tandem with the "Go Out" strategy, the Chinese government is promoting industry consolidation back home. Solar PV production appears to have peaked and many producers are exiting the market. Among them is the main operating unit of Suntech Power, once the world's premier solar panel manufacturer, which declared bankruptcy in March 2013 after Suntech's international holding company defaulted on a bond payment of over half a billion dollars.[105] The municipal government of Wuxi (in Jiangsu province) initially tried to help Suntech, but the company sold off its assets to a Hong Kong-based PV cell producer in the fall of 2013. Suntech's international holding company is also facing a lawsuit from bondholders in U.S. courts.[106]

Market corrections are beginning to occur in the wind sector as well. Due to an excessive number of turbine makers, the top-five turbine makers comprised just 65.3 percent of China's installed capacity in 2011, versus a top-five concentration of 86 percent in the United States that year. The Chinese Wind Energy Equipment Association recently predicted that industry consolidation is very likely as oversupply pressures grow, leaving as little as ten turbine makers in the market.[107] The success of Goldwind and the decline of its rival, Sinovel Wind Group Company Ltd., suggest that some market restructuring is already underway (see textbox below).

A related trend is the downstream integration of China's turbine makers into wind farm development. Goldwind's financial statements show the company generated 17.2 percent of its revenue from power generation and wind power services in the first half of 2014. Goldwind, as well as China Ming Yang Wind Power, China's second-leading turbine maker, has used the acquisition of land with good wind resources as a means to do business with developers and operators. These are essentially "swap deals," where the developer acquires rights to the wind farm site from the turbine maker in exchange for procuring that turbine maker's products.[108] For example, Goldwind formerly owned and operated the 109.5 MW Shady Oaks Wind Farm in Illinois, before selling it to a local U.S. power company in December 2012. [19]

A key challenge for Chinese companies is the fierce competition outside their home market. European and U.S. equipment makers, not to mention Japan's Mitsubishi and India's Suzlon, pursue sales across the world. According to the DOE, China-branded turbines occupied just 1.8 percent of the U.S. wind market during the boom year of 2012. The U.S. market was topped that year by GE (38.2 percent share), followed at some distance by leading European turbine makers Siemens, Vestas, Gamesa, and RePower. [20] The pattern was even more marked in 2013, a poor year for U.S. deployment, as GE accounted for 90.5 percent of the market. Despite some high-profile overseas projects, 94.1 percent of Goldwind's installed capacity in

2013 was deployed in China, where the company accounted for approximately one-quarter of the total capacity installed.[109] The company's only U.S. wind farm exceeding 10 MW was the aforementioned Shady Oaks project sold in 2012.[110]

For the time being, China's turbine companies are generating sales in emerging markets, which are less competitive but also less lucrative. Examples are Goldwind's recent shipments to Cuba (51 MW, December 2013),[111] Romania (50 MW, December 2013),[112] and Panama (215 MW, April 2014).[113] Goldwind is contemplating a wind turbine blade plant in Russia, in the context of recent improvements in China-Russia relations.[114] A central state-owned hydropower company, China Three Gorges Corp., owns a minority stake in Goldwind and is helping the company secure contracts in Pakistan and potentially Latin America, where Three Gorges has acquired power generating assets.[115] Meanwhile, Ming Yang is aggressively courting the Indian market, where it has established a subsidiary.[116]

Sinovel and Goldwind: Divergent Trajectories of China's Leading Turbine Makers

Sinovel Wind Group Co. was founded in 2004 with headquarters in Beijing. The company listed 10 percent of its shares on the Shanghai Stock Exchange in 2011, raising $1.4 billion. Officially in private hands, Sinovel has shareholders with strong ties to China's economic elite. [21] It has invested in R&D for offshore turbines, yet most of its sales revenue has come from low-cost, mature turbine technology that it licenses from Western companies and sells to state-owned grid operators in China.[117] Goldwind—private-owned and originated as a startup in western China in the late 1980s—has benefitted from Chinese government support as well, notably a $5.5 billion loan from China Development Bank.[118] [22]

Sinovel overtook Goldwind in 2010 as the top-ranked Chinese turbine company in terms of capacity installed. Since then, however, the two companies' fortunes have diverged. Goldwind has matured into the world's second-leading turbine producer, developing an innovative turbine drive technology—the permanent magnet direct-drive (PMDD)—together with Vensys, a small German company it acquired in 2008. Goldwind established a U.S. subsidiary in Chicago in 2010, and subsequently built a utility-scale wind farm in Shady Oaks, Illinois, with components sourced from several U.S. suppliers (Mortenson for project construction, LM Wind Power for blades, and Broadwind for towers).

Although Goldwind had a poor year in 2012, it quickly recovered and retained its position in the global top five.[119] The company further improved its sales in 2014, announcing in late October that its annual profit could quadruple.[120] Goldwind's share price on the Shanghai Stock Exchange has also outperformed its peers. [23]

Sinovel has seen a steep decline in sales turnover and has dropped outside the top-ten global turbine makers. In its 2013 earnings report, the company listed its 2013 net losses at RMB 3.4 billion ($605 million), compared to net income of RMB 2.8 billion ($424 million) just three years prior.[121] In April 2014, the Shanghai Stock Exchange formally suspended trading on Sinovel's corporate bonds because the company had reported negative net profits in two consecutive years.[122]

Sinovel is also fighting a series of legal battles that further strain its finances. [24] Of special interest to the United States is an intellectual property lawsuit involving the U.S. firm American Superconductor Corp. (AMSC). AMSC sued Sinovel in 2011 for stealing its turbine control codes and turning back its order consignments. AMSC originally filed the case in a Chinese court in 2011, to little effect. The case escalated in the ensuing months, with AMSC pursuing Sinovel for $1.2 billion in damages in China and $800 million in the United States.[123] In September 2011, an Austrian court convicted a former AMSC employee, a Croatian national, of selling trade secrets to Sinovel, and sentenced him to a year in prison.[124] In June 2013, the U.S. Department of Justice charged Sinovel with trade secret theft, alleging Sinovel executives and a former employee of AMSC conspired to obtain copyrighted information.[125] In July 2014, a U.S. magistrate judge ruled that Sinovel's attempt to present its subsidiary, Sinovel USA, as responsible for its activities in the United States was unfounded, and charged Sinovel along with two of its employees with stealing trade secrets. Back in China, the Supreme People's Court issued a February 2014 decision in favor of AMSC, ruling that the copyright infringement cases should be heard separately from the commercial arbitration claims. The next trial in the case is pending.[126]

Who Benefits from Clean Technology Trade?

The United States has registered a trade deficit in goods with China perennially since the 1980s. Whether clean energy trade contributes to or reduces this deficit is debatable. Based on calculations by the Pew Charitable

Trusts, the United States actually achieved a $1.63 billion "clean energy trade surplus" with China in 2011. Most of the surplus stemmed from U.S. exports of polysilicon and capital equipment to China for use in solar module assembly. Pew further calculates the United States achieved a $146 million bilateral trade surplus in the wind sector, owing mainly to blade materials exports (see Table 7). The wind and solar industries are characterized by a global division of labor and cooperative relationships among designers, suppliers, assemblers, and customers. These diverse stakeholders form networks that harmonize standards and promote incremental innovation. Within this international hierarchy, the United States and Europe enjoy an advantage in higher-margin and innovation-driven segments. [25] Chinese manufacturers, in turn, excel in labor-intensive production, economies of scale, and modular manufacturing.

Table 7. U.S.-China Clean Technology Trade, 2011 (Pew Estimates)

Solar Equipment

	US$ millions			Share (%)	
	Exports	Imports	Balance	Exports	Imports
Capital equipment	$ 2,200	$-	$ 2,200	59.1%	0.0%
Inverters	$10	$-	$10	0.3%	0.0%
Polysilicon	$ 684	$-	$684	18.4%	0.0%
Wafers	$ 300	$-	$300	8.1%	0.0%
Cells	$14	$151	$(137)	0.4%	5.4%
Modules	$12	$ 2,650	$ (2,638)	0.3%	94.6%
Materials	$ 500	$-	$500	13.4%	0.0%
Total	**$ 3,720**	**$ 2,801**	**$919**		

Wind Equipment

	US$ millions			Share (%)	
	Exports	Imports	Balance	Exports	Imports
Tower	$-	$ 196	$ (196)	0.0%	50.4%
Finished Turbine	$-	$26	$(26)	0.0%	6.7%
Blade Materials	$ 325	$7	$ 318	60.7%	1.8%
Electronics and Control	$73	$-	$73	13.6%	0.0%
Drive Train	$ 103	$ 160	$(57)	19.3%	41.1%
Rotor	$34	$-	$34	6.4%	0.0%
Total	**$ 535**	**$ 389**	**$ 146**		

Source: Pew Charitable Trusts, Advantage America: The U.S.-China Clean Energy Technology Trade Relationship in 2011 (Washington, DC, March 2013), pp. 15, 20.

However, these industry dynamics appear to be changing. China is persistently upgrading its component production to compete in segments where the United States enjoys an advantage. Data from the China Petroleum and Chemical Industry Federation, for example, shows China running a trade surplus with the world in solar PV wafers by selling at low prices (see Figure 5).

China is also improving its terms of trade in cell-grade silicon. Rapid growth in crystalline solar PV manufacturing caused temporary shortages of silicon in 2004–2007, resulting in higher prices on the world market. Because pure silicon feedstock production is very energy-intensive, prices were further inflated by the increase in global energy prices.[127] To establish a buffer against global silicon markets, China began to ramp up domestic production. The provincial government of Jiangsu, where most of China's solar panels are manufactured, set a target of 30,000 tons of polysilicon production by the end of 2011.[128] By year-end 2013, China ranked first in polysilicon output, with seven of the world's top-15 producers.[129] This coincided with severe overcapacities; China's polysilicon capacity utilization rates stood at less than 30 percent in the first half of 2013.[130]

Wind components are less traded, and until recently, most of this trade took place between Europe and the United States. Yet here too, China has started to take on market share. U.S. imports of blades and hubs exclusive to wind turbines have slumped recently, from $892 million in 2012 to $488 million in 2014. During this period, Brazil and Denmark, top exporters of this product to the United States, saw the value of their shipments decline. In contrast, China saw the value of its shipments increase, allowing it to raise its market share from 16.2 percent in 2012 to 36.7 percent last year. [26] China is also making headway on wind power generating sets. For all U.S. imports of this product between 1996 and 2011 ($12.6 billion), China accounted for only a 1.2 percent share. For all U.S. imports between 2012 and 2014 ($1.2 billion), China's share rose to 11.8 percent. While the United States has managed to expand its own exports of wind generating sets to the world, almost none of these exports are shipped to China. [27]

China has an incentive to move up the value chain because its advantage in labor-intensive production is diminishing. China's official statistics show average manufacturing wages outpacing economic growth (see Table 8). A separate indicator adjusted for inflation—the real wage index—also demonstrates that labor costs are on the uptick. Jiangsu province, China's solar PV manufacturing hub, is no exception.

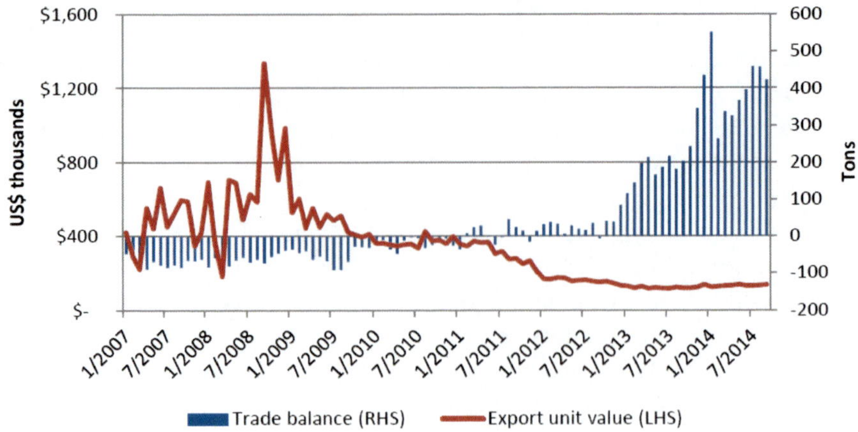

Note: "RHS" means right-hand side, "LHS" means left-hand side. *Source*: China Petroleum and Chemical Industry Federation, via CEIC.

Figure 5. China's Trade in Monocrystalline Silicon Wafers (>15.24 cm): Trade Balance and Export Unit Value. (Trade balance in tons; unit value in US$ thousands).

China's competitiveness in wind and solar manufacturing could be affected by rising costs for resource inputs, such as steel and aluminum. But resource prices have ebbed recently. Moreover, China derives considerable leverage from its near-monopoly over the world's rare earths supply. Pricing and availability concerns persist regarding neodymium (Nd) and dysprosium (Dy), the rare earth metals required for permanent magnets used in direct-drive turbines. A 2011 estimate found a single direct-drive turbine averaged 600 kilograms of permanent magnet material for every megawatt generated, totaling several hundred kilograms of rare earth materials.[131]

China began placing export restrictions on rare earths in 2009–2010, in the form of quotas, duties, and export licensing, arguing these were necessary to conserve natural resources and the environment.[132] In the ensuing years, the volume of U.S. rare earths imports from China decreased, even as the cost of those imports increased—a shift that favored China's rare earths exporters. A subsequent boost in production outside China did not completely resolve this problem.[133] Following a complaint brought forward by the United States, Japan, and the EU, a WTO panel in April 2014 deemed China's export restrictions on rare earths to be non-compliant with WTO rules. The panel found China did not place corresponding restrictions on domestic production and consumption.[134] Four months ahead of its May 2015 compliance deadline, China announced it would remove the rare earths export restrictions.[135]

Whether China will follow through and completely lift the restrictions is uncertain at this stage.

Table 8. Select Wage Data from China: Average Manufacturing Wages at Non-Private and Private Enterprises; Real Wage Index (Converted from RMB into US$)

	Average Manufacturing Wages						Real Wage Index	
	Non-private (US$)			Private (US$)			(Prior year = 100)	
	Annual	Hourly	yoy increase %	Annual	Hourly	yoy increase %	National	Jiangsu
2008	$ 4,075.5	$ 1.78		$ 2,410.6	$ 1.05		110.70	109.94
2009	$ 4,566.1	$ 2.00	12.0%	$ 2,530.7	$ 1.11	5.0%	112.61	112.96
2010	$ 5,476.5	$ 2.39	19.9%	$ 3,023.8	$ 1.32	19.5%	109.79	108.96
2011	$ 6,782.5	$ 2.96	23.8%	$ 3,804.6	$ 1.66	25.8%	108.64	108.61
2012	$ 7,513.8	$ 3.28	10.8%	$ 4,475.7	$ 1.96	17.6%	108.97	108.52
2013	$ 8,846.8	$ 3.87	17.7%	$ 5,239.2	$ 2.29	17.1%	107.31	110.04

Note: Based on 2,288 working hours per year (44-hour workweek) and nominal exchange rates at year end. "Yoy" refers to year-on-year. *Source*: China National Bureau of Statistics, via CEIC.

The Development of New Technologies

Technological advances play a vital role in making wind and solar deployment competitive in the energy market. [28] R&D is also an important dimension of the U.S.-China clean energy relationship. Below is a brief assessment of how the two countries are promoting the use of better technologies.

Grid Infrastructure Technology

Grid connectivity is a chief impediment to wind and solar deployment, particularly in countries with expansive territories like the United States and China. Various solutions are under development to optimize grid operations:

- *Ultra high-voltage (UHV) transmission technology.* UHV transmission lines minimize transmission losses over long distances. China already has seven UHV lines in operation, the most of any country. Most long-distance transmission lines in the United States operate at 230 kilovolts (kV), 345 kV, or 400 kV; in China, they mostly operate at 800 kV. In April 2014, the Chinese government approved the construction of another line operating at 1,000 kV between rural Anhui province and the cities of Nanjing and Shanghai. The NDRC has ambitious plans for as many as 12 inter-regional UHV transmission corridors spanning the country.[136] These plans are part of China's West-East Electricity Transfer project, which aims to transport electricity generated in the scarcely populated western regions to population centers further east. [29]

- *Smart grids.* Smart grids gather and act on information about the behaviors of electricity suppliers and consumers in an automated fashion. They can respond flexibly to intermittency problems associated with wind and solar power, as well as prevent overproduction of conventional baseload power (coal, nuclear, and gas). Installing smart grids involves three phases: (1) smart meters, (2) distribution automation, and (3) information integration. Of the $45 billion China pledged to spend on smart grid infrastructure under the 12th Five-Year Plan, the majority has been allocated toward installing new meters in homes and businesses.[137] China outspent the United States for the first time on smart grid technology in 2013 with investments of $4.3 billion, nearly one-third of the global total. Experts predict China's current wave of smart meter installations to end in 2017, at which point the focus will turn to distribution automation and information integration. At this rate, China's smart grids could soon achieve parity with the United States.[138]
- *Energy storage technology.* Advanced battery systems are being developed to store wind and solar electricity generated during off-peak periods, to allow for timely deployment during periods of peak demand. Alternatively, stored energy could be used for heating or off-grid power applications in homes and businesses. IHS Technology predicts grid-connected global energy storage capacity will rise exponentially, from 340 MW in 2013 to 40 GW by 2022.[139] So far, the technology remains at a pilot stage. In September 2014, the California Public Utilities Commission set a goal of 1.3 GW of energy storage to support the state's power grid by 2020.[140] China's battery energy storage sector, still in its infancy, is oriented toward renewable energy integration (see Figure 6).

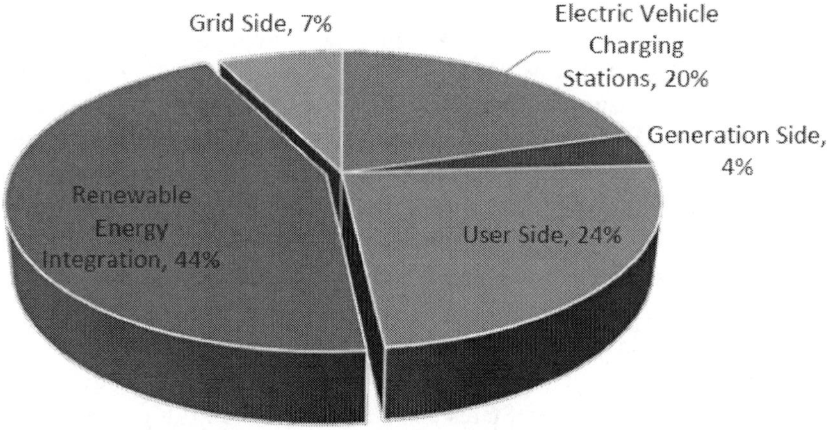

Source: Adapted from CNESA ES Project Database, 2014, via Electrical Energy Storage Magazine. http://ees-magazine.com/energystorage-in-china/.

Figure 6. Battery Energy Storage in China by Application. (Total operational = 54 megawatts).

Wind and Solar Equipment Components

Wind and solar components undergo constant innovation, both incremental and disruptive. A wind turbine contains no fewer than 8,000 components, so there is ample room for improvements. Innovations occur foremost in (1) the rotor, which consists of the blades, hub and spinner, and (2) the nacelle, which contains most of the components. A key distinction is between the popular "shaft-gear" nacelle and the more sophisticated PMDD system. Solar modules use fewer components than wind turbines, and thus offer less opportunity for modifications. What is similar to the turbine market is that one type of technology—in this case, crystalline silicon—holds greater market share, while another—thin film PV—is more advanced but less widely deployed (see textbox below for details).

China's Goldwind is the principal producer of PMDD technology. The company owns Vensys, a small German firm that helped pioneer PMDD. China has one of the highest concentrations of PMDD turbines in the world—at least 25 percent, according to estimates.[141] In the United States, the corresponding market is significantly smaller; since 2008, less than 5 percent of the country's new turbine installations used PMDD.[142] GE Wind's 1.5/1.6+ MW shaft-gear turbines were the United States' most-popular turbine in 2012, totaling 33 percent of all capacity installed.[143] Direct-drive systems are well-suited to giant turbines, so the growing pains of the offshore wind segment—especially in the United States—could hamper their wider application.

Turbine makers in China are also focusing on designing longer rotor blades to accommodate the country's low wind speeds, adverse weather conditions, and high altitude areas. Goldwind was among the first Chinese firms to develop rotor blades for this purpose. These blades are longer than those for typical utility turbines, equipped with protection against lightning and radiation, and thermally insulated. Goldwind in December 2011 completed its first high altitude wind farm project in the Yunnan-Guizhou Plateau, using 33 sets of 1.5 MW turbines.[144]

In contrast to the wind turbine sector, where China has spearheaded the adoption of direct-drive technology, cutting-edge solar PV technology is more widely deployed in the United States than China. U.S.-based First Solar is the world's number-one thin film solar manufacturer, and the only thin film producer among the world's top-ten producers of solar PV. Using cadmium telluride technology, First Solar managed to bring the cost of thin film below $1 per watt for the first time in February 2009, and in 2013 achieved an efficiency rate comparable to crystalline silicon PV.[145] In China, on the other hand, virtually all production and installation revolves around crystalline silicon PV. Slow adoption of thin film PV is in part a consequence of China's modular supply chains, which make it difficult to adopt new technologies that cut out established suppliers.

The Advantages and Limitations of Key Wind and Solar Technologies

Wind Turbines

Rotor blades: Within the rotor, the quality and length of a turbine's blades are crucial to determining power capacity. Turbine manufacturers are increasing blade size to accommodate bigger turbines and improve electricity generation under low-wind-speed conditions. Joint research by GE, Virginia Tech, and the National Renewable Energy Laboratory (NREL) suggests next-generation blades will need to grow by up to 50 percent in length.[146] Currently, blades average between 30 and 50 meters, and on "super-turbines" upwards of 7 MW, can reach 75 meters in length.[147] Turbine firms have also attempted to reduce the blade cost relative to other components. Rotor blades make up around one-fifth of the total turbine cost.[148]

The industry is experimenting with turbine blades that combine light weight with durability. Standard practice is to bolt the blades on a cast-iron hub, one of the heaviest components in a wind turbine (typically weighing eight to ten tons). New material options undergoing laboratory testing are carbon fiber, composites, and fiberglass. In the United States, GE, Virginia Tech, and NREL are jointly experimenting with architectural fabrics wrapped around a metal frame.[149]

"Shaft-gear" vs. PMDD systems: In traditional shaft-gear systems, the blades spin on a shaft connected through a gearbox to the generator. In recent years, PMDD technology using permanent magnets has emerged as an attractive alternative that could make the shaft-gear system obsolete. Direct-drive generators are lighter, smaller, and more power-efficient. With fewer moving parts, they become economical in the long run due to low operations and maintenance (O&M) costs.[150] Perhaps most important, PMDD technology is optimal for the giant turbines currently entering the market, as such turbines are particularly difficult to service when they break down.

Direct-drive systems also come with disadvantages. They carry a heftier price tag, forcing wind farm developers to incur high upfront capital costs. The principal cost driver of direct-drive technology is its narrow, specialized supply chain. With fewer modular components, direct-drive systems require significant in-house capabilities to manufacture, and offer less room for competitive outsourcing. In addition, the price of the rare earth metals used in permanent magnets has fluctuated over time and is susceptible to China's restrictions on production and exports. An added disadvantage is the proprietary nature of direct-drive technology, which makes it difficult for original equipment manufacturers to license.

Solar PV

Thin-film PV: Thin-film, still less than 10 percent of the global solar market, is significantly cheaper than crystalline silicon. Thin-film panels require less semiconducting material, and production can be done in one casting. The panels are lighter as well, which makes them easier to transport. Recent lab tests show thin-film is quickly catching up to crystalline silicon PV in efficiency rates (i.e., capacity utilization of a solar module). An additional benefit is that most thin-film PV deployed uses tellurium, gallium, and indium, which do not have the high costs of purification associated with crystalline silicon PV.

The disadvantage of thin-film PV is that, due to lower efficiencies, surface area requirements are higher. Also, some non-silicon input materials are rare earths (e.g., tellurium) and others (e.g., cadmium) can be a hazard to human health.[151]

Multi-junction silicon PV: Beyond traditional silicon-based and thin-film PV, other technologies have the potential to enter the market and alter the status quo. A leading emerging technology is multi-junction PV, which uses multiple semiconducting materials that can absorb a broader range of wavelengths, improving the cell's sunlight to electrical energy conversion efficiency. Researchers in France and Germany recently achieved a 46 percent efficiency rate using this technology—the highest rate ever recorded for a solar panel.[152] The challenge is that multi-junction PV requires additional frames that can be adjusted to follow the path of the sun, adding a layer of complexity and cost to the manufacturing process.[153]

Carbon-based solar cells: New findings are also paving the way for carbon-based solar cells, which are much cheaper than silicon cells.[154] Stanford scientists recently built the first "all-carbon" solar cell, while Yale scientists have invented a method to use smooth carbon nanotube films to create hybrid carbon/silicon cells.[155] The problem— not unlike with thin-film PV—is that carbon-based cells are inferior to crystalline silicon PV in terms of conversion efficiency.[156]

SECTION 4: ENERGY SECURITY AND THE ENVIRONMENT

U.S. and Chinese Energy Security

Import Dependence and Energy Demand

Energy security [30] can be a powerful incentive to seek new sources of energy. Fossil fuel importers share concerns about the security of supply from distant and unstable regions, as well as the short-term volatility and long-term rise in energy prices. Wind and solar power have the potential to diversify a country's energy mix without depleting domestic resource endowments or necessitating energy imports.

At present, China has greater reason to be concerned about its energy security than the United States.

- China in 2013 became the world's largest net importer of oil (domestic production minus domestic consumption). Bolstered by domestic shale oil and gas production, U.S. oil imports hit a 20-year low in 2012, and U.S. reliance on oil imports fell to 39 percent, compared with 59 percent in 2007. China's rate of net oil import dependence reached a new high of 56 percent in 2012, up from just 26 percent a decade earlier.[157]

- In September 2014, the EIA predicted the United States in 2040 will only source one-third of its crude oil from international markets. China, on the other hand, will import nearly three-quarters to satisfy its needs. By that point, China could also overtake the United States as the world's top oil consumer.[158]
- The situation is similar with natural gas. Although China accounts for a smaller share of global natural gas consumption (4.3 percent in 2012) than the United States (21.4 percent), its consumption doubled in 2007– 2012. Natural gas import dependence could compound China's energy insecurity.[159]

In the United States, energy demand is not placing a strain on supply. U.S. oil consumption fell by 11 percent in 2007–2012, owing chiefly to an economic downturn, tougher fuel economy standards, and reduced vehicle use by the younger generation.[160] For similar reasons, electricity generated in the United States from 2009 to 2013 totaled less than in the preceding five years combined.[161] China, on the other hand, witnessed a 37 percent increase in oil use and a 34 percent rise in electricity use in 2007–2012.[162] Behind these trends is a combination of rapid GDP growth and, in spite of recent improvements, high energy intensity by international standards (see Figure 7).

As of 2012, the industrial sector accounted for over 70 percent of China's energy consumption (see Figure 8). The corresponding figure in the United States was 34 percent.[163] Several heavy industries (glass, cement, aluminum) suffer from overcapacity and add limited value to China's economy. Recently, these industries have migrated from eastern China into the interior and western regions in search of fiscal incentives and cheaper energy costs. [31] Concurrently, China's energy consumption is booming in the transport and housing segments. China's urbanization rate (officially 53 percent in 2013) is nearly 30 percentage points lower than in the United States.[164] New buildings are appearing across the country, combined with greater energy intensity per square meter as living standards improve. China's per capita ownership of vehicles is still moderate by U.S. standards, but it already became the world's top-selling vehicle market in 2010.[165]

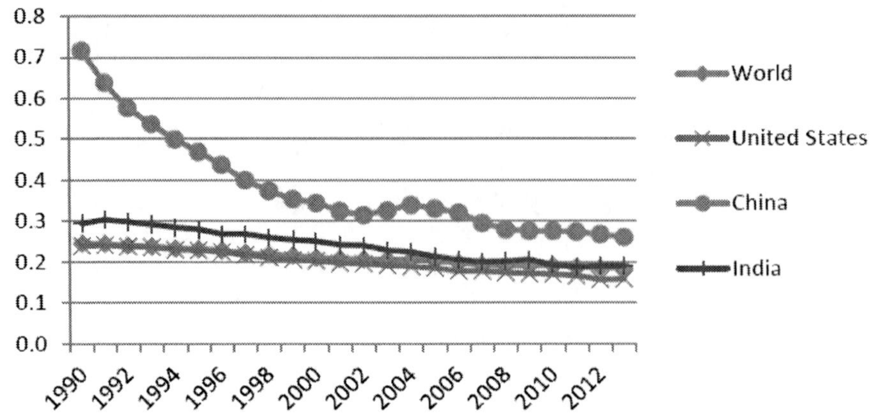

Source: Enerdata.

Figure 7. Energy Intensity of GDP at Constant Purchasing Power Parities, 1990–2013. (Kilograms of oil equivalent per 2005 dollar).

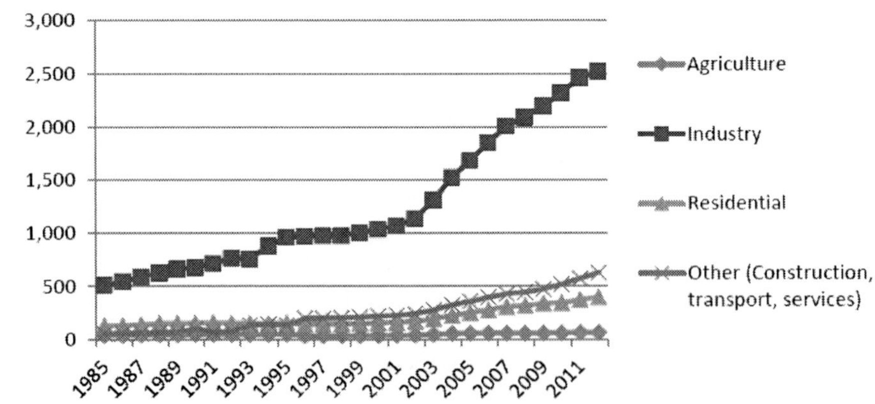

Note: "SCE" refers to standard coal equivalent.
Source: China National Bureau of Statistics, via CEIC.

Figure 8. China Energy Consumption by Sector, 1985-2012. (Standard coal equivalent tons millions).

To an extent, China's authoritarian government has the ability to reduce energy demand by fiat. Under the 11th Five-Year Plan, China introduced performance metrics into its cadre evaluation system, in order to hold local officials accountable for reducing emissions and energy intensity.[166] During the duration of that five-year plan, China achieved energy savings equivalent to 500 million tons of oil, thanks mainly to closures of small power plants and mines, reductions in energy use at the top-1000 most polluting enterprises, and improved energy efficiency in buildings.[167] During the 12th Five-Year Plan, China has continued these initiatives; for example, by expanding the Top-1000 program to 10,000 companies.[168]

There has also been progress on energy prices and taxes. The NDRC loosened its price controls on refined petroleum products in 2013. These prices are now adjusted every ten days in line with international market trends. Previously, prices were adjusted every 22 days, and only if the price moved by more than 4 percent.[169] Last December, China also introduced a price-based resource tax on coal consumption, taking advantage of a lull in energy prices. A price-based tax is superior to a volume-based tax because it tracks markets more closely.[170] Based on citations in Chinese-language policy journals over the last few years, there appears to be some interest in instituting a carbon tax, though there are no official indications to date.[171]

In spite of these progressive moves, the fact is demand-side measures are easier to design than execute. Central-level officials contend with opposition from local governments in industry-heavy provinces who are preoccupied with sustaining jobs and fiscal revenue.[172] Chinese households also spend an inordinate amount of their income on basic necessities. In light of rising inequality and a weak social safety net, the government relies on cheap energy to contain inflation and subsidize real incomes.[173]

Low-Carbon Energy Sources in China's Energy Supply

On the supply side of the energy sector, there are numerous means by which China can improve its energy security. Smarter procurement and use of petroleum [32] and coal [33] is one solution. Equally if not more important are low-carbon energy initiatives. These are not limited to wind and solar; hydropower, natural gas, and nuclear power can all deliver affordable electricity at a steady rate. Below is a brief description of these alternatives and their current application in China.

Hydropower

China ranks first in installed hydropower capacity (61.4 million tons of oil equivalent (Mtoe)), well ahead of second-ranked Brazil (36.9 Mtoe).[174] Last decade, the country's hydro capacity registered the largest GW gains of any renewable energy source. The government by 2030 plans to add 1,212 GW of hydro capacity, equivalent to six times India's current installed power capacity.[175]

Realizing this goal will be difficult. Water is essential to irrigation, industry, and households. Over the period 2003– 2010, China's national renewable-water resources per capita averaged 2,000 cubic meters, just above the stress level of 1,700 cubic meters.[176] Due to uneven distribution, 11 provinces are already water scarce.[177] Faced with water scarcity in the arid northeast, the central government has allocated $65 billion dollars toward the South-North Water Transfer Project, which aims to divert 45 cubic kilometers of water a year northward.[178] At the same time, the Three Gorges Dam and other hydro projects have altered flood patterns and displaced populations, with far-reaching social implications that the Chinese government is only starting to confront.[179] Hydropower has also caused tensions between China and its southern neighbors, as China owns headwaters to at least 10 major rivers.[180]

Natural Gas

Natural gas is a clean source of energy, and compared to coal can be applied flexibly. In addition to electricity generation, it can be used for heating and for transport fuel (converted gasoline, liquefied natural gas (LNG), and compressed natural gas).[181] An influential report by the NDRC's Energy Research Institute, published in 2012, advocates raising gas imports from abroad, especially into southeast China, where it could relieve the bottlenecks that plague rail-based coal shipments from the northwest of the country.[182]

However, some Chinese policymakers are wary that China could become too reliant on foreign gas supplies. The ratio of imports to consumption is currently around 25 percent.[183] China holds the world's most abundant reserves of unconventional gas, so ramping up domestic production is a viable option. The decision by China's state-owned oil firms to invest billions of dollars in North American exploration projects is viewed by some as a strategy to "road-test" technologies for use back home.[184] China's first hydraulic fracturing (fracking) projects are underway, led by China National Petroleum Corp. in the northeast.[185]

Analysts at Bloomberg, however, see China's targets of producing 80 billion cubic meters of shale gas by 2020 as unrealistic, and predict output to approach a mere 18 billion cubic meters.[186] China's deposits are deeper and more scattered than in the United States.[187] There are also environmental concerns associated with fracking, [34] including excessive water consumption, potential groundwater contamination, waste water disposal, and the possibility of earthquakes [35] near fracking sites.[188,] [36] In addition, shale projects in China have been subject to excessive government interference.

China began auctioning domestic shale gas projects in June 2011,[189] with a third bidding round expected sometime this year. An energy expert at *Forbes* magazine claims the bidding decisions are biased:

> China's auction of shale blocks is a somewhat strange business. The first shale sale in June 2011 was largely a staged play among the six state-run bidders. The second round held in 2012 was open to private investors, but shortlisted candidates included the bizarre choices of a home-appliance company and a hardware firm that mainly sells hand tools. Now the repeatedly delayed third round is expected to take place, but things are unlikely to get any better for private investors hoping to benefit from the country's future shale gas boom.[190]

Nuclear Energy

Nuclear energy, which currently accounts for just 2 percent of China's energy mix, is a common substitute for coal in baseload power generation. In the aftermath of the 2011 Fukushima Daiichi disaster in Japan, governments around the world reassessed their nuclear power strategy. Germany decided to remove all nuclear power by 2022.[191] The United States and China, on the other hand, did not abandon their plans. Indeed, the U.S. Nuclear Regulatory Commission in late 2011 approved construction of four new reactors in South Carolina and Georgia, the first such approvals in three decades.[192]

After conducting a safety review of all nuclear plants in operation and under construction, China in mid-2012 opened the door for new nuclear power plant approvals as well. By then, China was building almost half of new global nuclear capacity, over 33 GW, with plans to reach 40 GW of capacity by 2015 and at least 70 GW by 2020.[193] China now has 22 nuclear power reactors in operation, 26 under construction, and more in the pipeline.[194]

China is using cutting-edge AP1000 reactors designed by the U.S. company Westinghouse and other foreign suppliers.[195] Westinghouse technology is being transferred to China in conjunction with the U.S.-China Peaceful Uses of Nuclear Technology (PUNT) Agreement. [196] The World Nuclear Association commends China's "unprecedented eagerness to achieve the world's best standards in nuclear safety."[197] China conducts at least one external safety review a year at each of its nuclear plants, and is beginning to collaborate with nuclear experts in Japan and South Korea.[198]

China could still run the risk of building out its nuclear capacity too quickly. There is a direct economic incentive for it to pursue nuclear power, as much of the reactor design and construction, as well as component production, is done domestically. China's political leaders have recently shown an interest in exporting the country's indigenous nuclear technology—a difficult sell, since it has not been installed on a large scale within China.[199] Also worth considering is China's susceptibility to earthquakes, and a poor safety record in other industries, such as high-speed rail.[200]

A study by the Lawrence Berkeley National Laboratory (LBNL) provides useful projections for China's future energy supply. The study presumes China will continue to shut down small, inefficient coal-fired power plants and also succeed in generating more power from renewable sources. Notably, the study sees greater potential for wind and nuclear than for natural gas, hydro, and solar (see Table 9).

Table 9. China's Power Generation Shares by Technology under Different Scenarios (to 2050) (Lawrence Berkeley National Laboratory)

	2005	Continued improvement		Accelerated improvement	
		2030	2050	2030	2050
Wind power (%)	0	6	13	10	16
Nuclear power (%)	2	13	25	19	54
Natural gas combined cycle power plants (%)	1	2	2	4	2
Hydropower (%)	15	12	12	17	16
Oil fired units (%)	2	0	0	0	0
Biomass and other renewable (%)	0	1	1	1	1
Solar (%)	0	1	1	1	1
Coal 100 MW (%)	21	0	0	0	0
Coal 100–200 MW (%)	11	0	0	0	0
Coal 200–300 MW subcritical units (%)	9	0	0	0	0
Coal 300–600 MW subcritical units (%)	35	1	0	0	0
Coal 600–1000 MW supercritical units (%)	2	22	8	19	0
Coal 4 1000 MW ultra-supercritical units (%)	0	44	41	30	9

Source: Nan Zhou et al., "China's Energy Emissions Outlook to 2050: Perspectives from Bottom-Up Energy End-Use Model," *Energy Policy* 53 (2013): 60.

The looming question with all low-carbon energy sources is how compatible they are with intermittent wind and solar power. Gas-fired power plants are relatively cheap to build and flexible to use, and thus may constitute a better baseload power source than hydro and nuclear. At the same time, grid operators still face the basic dilemma that wind and solar power is not available on demand. Thus, even if their levelized cost declines relative to other low-carbon sources, wind and solar will remain at a disadvantage compared to more reliable sources that consistently meet peak demand.

The Environment and Climate Change Mitigation

China and the United States are the world's worst carbon emitters, but neither country has agreed to binding, multilateral emissions cuts. (The United States has not ratified the Kyoto Protocol, while China as a non-Annex B signatory does not have binding targets.) After failing to reach a consensus at the Copenhagen Climate Summit in 2009, Beijing and Washington have begun to make progress at the bilateral level.

- In the summer of 2013, President Obama and President Xi forged a deal at the Sunnylands Summit to cut HFC emissions. The United States had been pushing for years to get Chinese support for a global phase-down of HFCs under the Montreal Protocol Treaty on Substances that Deplete the Ozone Layer, originally signed in 1987 to curb the use of ozone-depleting chemicals. [201] Melanie Hart of the Center for American Progress estimates that amending the original treaty could avoid half a degree Celsius of warming by the end of the 21st century.[202]
- At the Strategic and Economic Dialogue talks a month after the Sunnylands Summit, the two countries agreed to a five-part plan to cut carbon emissions, with an emphasis on efficient heavy vehicles and coal-fired plants.[203]
- The Asia-Pacific Economic Cooperation (APEC) summit in Beijing in November 2014 saw further progress. President Obama declared targets for the United States to double the average pace of carbon reductions after 2020, for an overall reduction of between 26 percent and 28 percent by 2025, compared with 2005 levels.[204] This followed the publication in 2013 of *The President's Climate Action Plan*, which details the steps the Obama Administration will take to achieve its

commitment to reduce greenhouse gas emission by 17 percent from 2005 levels by 2020.[205] [37] China, in turn, announced targets to peak CO2 emissions around 2030, with the intention to try to peak early, and reiterated its goal to raise the share of non-fossil fuels to 20 percent of primary energy consumption by 2030.[206] The agreement could pave the way for a new international climate deal at the 2015 United Nations Climate Change Conference in Paris.

China is also pursuing intensity-based targets.

- *Carbon intensity*. [38] China aims to cut emissions per unit of GDP by 40 to 45 percent by 2020 from 2005 levels, a target it announced at the Copenhagen Climate Change Conference in 2009. In the 12th Five-Year Plan, China committed to reducing the carbon intensity per unit of GDP by 17 percent from 2010 to 2015, implying approximately 3.5 percent in reductions each year.[207] No up-to-date data has been made available by the government to track progress on this front. China's chief climate negotiator claimed in early 2013 that carbon intensity fell by 3.5 percent in 2012.[208] It is difficult to verify this claim, and subsequent announcements have not been made for the years 2013 and 2014. An annual energy review by British Petroleum claims China's carbon emissions growth in 2013 was half of average levels over the past decade. However, the BP data also suggests it will be difficult for China to meet its 2020 target.[209]
- *Energy intensity*. In the 11th Five-Year Plan, China for the first time committed to reducing energy intensity. China claims to have reduced energy intensity per unit of GDP by 19.1 percent during the Plan, just short of the 20 percent target.[210] Independent experts at LBNL, Yale and Columbia universities, and the International Energy Agency have confirmed China's official results.[211] The 12th Five-Year Plan subsequently set a lower target of 16 percent reductions. A May 2014 report by the International Partnership for Energy Efficiency Cooperation claims China is on track to meet this target.[212]

As part of its carbon reduction strategy, the Chinese government in 2013 launched three carbon trading platforms in Shenzhen, Beijing, and Shanghai. Analysts at BNEF predict China's cap-and-trade program will rank second only to Europe's by the end of this year.[213] There are currently seven operating pilot programs, and details on the national plan are expected to be released in

2015. The California Air Resources Board (CARB), as well as the Massachusetts Institute of Technology, provided resources and advice during the formulation process of Shenzhen's pilot program. [39] The June 2013 launch of the Shenzhen pilot was accompanied by a new order calling on firms in heavy-polluting industries to cut their emissions by 30 percent by 2017.[214] About 95 percent of state and provincially controlled firms in China now conduct carbon emissions monitoring.[215]

Air pollution makes emissions reduction that much more pressing for the Chinese government. While air quality has improved markedly in the United States since the 1970 Clean Air Act, airborne particulate matter is now a leading cause of social unrest and healthcare problems in China.[216] Air pollution in Beijing has reached such poor levels that the state-run news media in 2013 broke from its customary downplay of environmental concerns to report the true severity of the problem.[217] At the 12th National Party Congress in March that year, outgoing premier Wen Jiabao announced the government would start to monitor fine particulate matter (PM2.5) in provincial capitals and megacities.[218] However, ahead of the November 2014 meeting of Asia-Pacific leaders in Beijing, rumors circulated that Chinese authorities had asked sites and apps to stop using air pollution data provided by the U.S. embassy, a timely source popular among Chinese citizens.[219]

IMPLICATIONS FOR THE UNITED STATES

In all likelihood, wind and solar power will make only a marginal contribution to global energy security and the environment in the next 10 to 15 years. Flaws inherent to each sector, combined with external constraints, will take time to address. Nevertheless, wind and solar continue to play an important role in manufacturing, trade, services, and R&D activities. On a level playing field, China and the United States stand to benefit from mutual engagement. Trends in China's wind and solar sectors have important implications for the United States.

- *The right balance between trade enforcement and bilateral cooperation.* Some U.S. manufacturers argue trade remedies against Chinese clean technology help level the playing field and send a message that the United States will not tolerate violations of WTO commitments. Others argue that trade remedies impede the functioning of global supply chains and make it more difficult for clean technology to compete against cheap conventional fuel sources. As U.S.-China cooperation deepens on issues such as climate change mitigation and clean energy development, disagreements over the proper use of trade remedies could intensify.

- *The risk of importing lower-quality components from China.* Chinese renewable energy equipment imported into the United States can be both affordable and good quality. Notwithstanding these benefits, there have been several instances of quality lapses. Equipment acquired from China may be cheaper but ultimately less economical if it leads to equipment failure and low efficiency.[220] The risk is especially great for utility-scale projects where cost is a major factor. More research could be done to compare Chinese and U.S. products over a power-generating project's lifetime.
- *Tougher competition in high-margin segments.* Once reliant on foreign suppliers for clean energy components, China has become fairly self-sufficient, to the extent that it is now exporting sophisticated components like silicon wafers and wind power generating sets. This adds to the leverage the Chinese government derives from a massive domestic market. Upgrading can have positive effects when it strengthens supply chains and spurs innovation. In China's case, however, wind and solar manufacturing are also part and parcel of an aggressive industrial policy that disadvantages U.S. firms.
- *The emergence of China's middle-income class.* Beginning with the 11th Five-Year Plan, the Chinese government has aimed to rebalance the economy away from fixed investment and export-led manufacturing toward domestic consumption and services. While rebalancing has been a slow and incomplete process, disposable income is growing and a new middle-income class is emerging. Ordinary Chinese could become more willing to pay for wind and solar energy and high-quality U.S. equipment if it contributes to a cleaner environment and better quality of life.
- *Carbon trading and a post-Kyoto climate agreement.* The United States and China's joint commitment to cap carbon emissions by 2030 could increase the potential for developing and developed countries to sign onto multilateral emissions reductions at the United Nations Climate Change Conference in December 2015. If a comprehensive agreement is reached, it could revitalize emissions trading schemes that generate carbon credits from wind and solar projects.

APPENDIX

Tables

Table A1. Cumulative Wind Energy Capacity Installed by Country, 2012–2013 (GW; share %)

		GW			Share (%)		
		End 2012	New 2013	End 2013	End 2012	New 2013	End 2013
1	China	75.3	16.1	91.4	26.6%	46.1%	28.7%
2	US	60.0	1.1	61.1	21.2%	3.1%	19.2%
3	Germany	31.3	3.0	34.3	11.0%	8.5%	10.8%
4	Spain	22.8	0.2	23.0	8.0%	0.5%	7.2%
5	India	18.4	1.7	20.2	6.5%	5.0%	6.3%
6	UK	8.6	1.9	10.5	3.1%	5.4%	3.3%
7	Italy	8.1	0.4	8.6	2.9%	1.2%	2.7%
8	France	7.6	0.6	8.3	2.7%	1.8%	2.6%
9	Canada	6.2	1.6	7.8	2.2%	4.6%	2.5%
10	Denmark	4.2	0.6	4.8	1.5%	1.7%	1.5%
	Top10	242.6	27.2	269.8	85.7%	77.9%	84.8%
	RoW	40.6	7.7	48.3	14.3%	22.1%	15.2%
	World	**283.2**	**34.9**	**318.1**	**100.0%**	**100.0%**	**100.0%**

Note: "RoW" refers to rest of world.

Source: Global Wind Energy Council, *Global Wind Report – Annual Market Update 2013* (April 2014). *http://www.gwec.net/wp-content/uploads/2014/04/GWEC-Global-Wind-Report_9-April-2014.pdf.*

Table A2. Cumulative Solar Energy Capacity Installed by Country, 2012–2013
(GW; share %)

		GW			Share (%)		
		End 2012	New 2013	End 2013	End 2012	New 2013	End 2013
1	Germany	32.6	3.3	35.9	32.6%	8.5%	25.8%
2	China	7.0	12.9	19.9	7.0%	33.1%	14.3%
3	Italy	16.4	1.5	17.6	16.4%	3.8%	12.7%
4	Japan	6.6	6.9	13.6	6.6%	17.7%	9.8%
5	US	7.2	4.8	12.1	7.2%	12.3%	8.7%
6	Spain	5.4	0.2	5.6	5.4%	0.5%	4.0%
7	France	4.0	0.6	4.6	4.0%	1.5%	3.3%
8	UK	1.8	1.5	3.3	1.8%	3.8%	2.4%
9	Australia	2.4	0.8	3.3	2.4%	2.1%	2.4%
10	Belgium	2.7	0.2	3.0	2.7%	0.5%	2.2%
	Top10	86.1	32.7	118.9	86.1%	83.8%	85.5%
	RoW	13.8	6.5	20.2	13.8%	16.7%	14.5%
	World	**100.0**	**39.0**	**139.0**	**100.0%**	**100.0%**	**100.0%**

Note: "RoW" refers to rest of world.

Source: REN21, *Renewables 2014 Global Status Report* (2014).

http://www.ren21.net/portals/0/documents/resources/gsr/2014/gsr2014_full%20report_low%20res.pdf

Table A3. Definition of the Main Solar PV and Wind Turbine Components

Wind Components

Rotor	The rotor usually consists of three blades, a hub, and a spinner. It creates the mechanical energy needed to generate electricity.
Blades	Blades are designed to generate lift, similar to airplane wings, causing the rotor to turn.
Hub	The blades are bolted onto the hub, which is designed to absorb vibrations and steady the blades. The hub—usually made of cast iron— is one of the heaviest components of a turbine, weighing 8-10 tons for a 2-MW turbine.
Spinner/Nose cone	The spinner or nose cone covers the hub and can protect the hub from the outside environment.
Nacelle	Made of fiberglass, the nacelle sits atop the tower and houses most of a turbine's 8,000 components. It protects the internal components from the outside environment and provides support for the main frame.
Gear box	Connecting the low-speed shaft to high-speed shaft, the gear box increases the rotational speed to 1,000-1,800 revolutions per minute—the speed most generators require to produce electricity. This component is also heavy and costly. Direct-drive turbines do not require a gear box.
Generator	The generator converts mechanical energy produced by the rotor into electricity.
Tower	Supporting the structure of a turbine, the tower is made of tubular steel, concrete, or steel lattice. Taller towers generate more electricity due to increases in wind speed.

Source: American Wind Energy Association, "Anatomy of a Wind Turbine." *http://www.awea.org/Resources/Content.aspx?ItemNumber=5083&RDtoken=29819&userID=4379*; U.S. DOE Office of Energy Efficiency and Renewable Energy, "The Inside of a Wind Turbine." *http://energy.gov/eere/wind/inside-wind-turbine-0*.

Solar Components

Polysilicon	Polysilicon is the material from sand used to build semiconductors that convert sunlight into electricity.
Wafer	A wafer is a highly precise, thinly cut piece of silicon. The process uses highly sophisticated manufacturing techniques to form the bases of a solar cell.
Cell	Solar cells are made through chemically treating the surface of the wafer to focus electrons when exposed to sunlight.
Module	The final step in the manufacturing process is wiring cells together with a protective outer layer to create a module, the finished solar panel.
Monocrystalline	A type of solar cell that uses silicon pulled into a single crystal.
Polycrystalline (Multicrystalline)	A type of solar cell where several crystals form in a mold. These cells have a "metal flake effect" appearance.

Source: SolarWorld, "Solar Energy 101." http://www.solarworld-usa.com/solar-101/solar-glossary.

Table A4. Top-10 Solar PV Module Suppliers by Shipments, 2011–2013

	2011		2012		2013	
1	First Solar	US	Yingli Green Energy	China	Yingli Green Energy	China
2	Suntech	China	First Solar	US	Trina Solar	China
3	Yingli Green Energy	China	Suntech	China	Sharp Solar	Japan
4	Trina Solar	China	Trina Solar	China	Canadian Solar	Canada
5	Canadian Solar	Canada	Canadian Solar	Canada	Jinko Solar	China
6	Sharp Solar	Japan	Sharp Solar	Japan	ReneSola	China
7	Hanwha SolarOne	China	Jinko Solar	China	First Solar	US
8	Jinko Solar	China	JA Solar	China	Hanwha SolarOne	China
9	LDK Solar	China	SunPower	US	Kyocera	Japan
10	SolarWorld	Germany	Hanwha SolarOne	China	JA Solar	China

Source: Sile Mc Mahon, "First Solar and Suntech Led 2011's Module Manufacturer Rankings, Says Lux Research Report," PV Tech, March 12, 2012. http://www.pv-tech.org/news/first_solar_and_suntech_led_2011s_module_manufacturer_rankings_says_lux_res; Ray Lian, "Top 10 PV Module Suppliers in 2012," PV Tech, January 28, 2013. http://www.pv-tech.org/guest_blog/top_10_pv_module_ suppliers_in_2012; and Ray Lian, "Top 10 PV Module Suppliers in 2013," PV Tech, January 8, 2014. http://www.pv-tech.org/guest_blog/top_10_pv_ module_suppliers_in_2013.

Table A5. Top-10 Wind Turbine Producers by Megawatts Installed, 2013

	Company	Country	Megawatts	Share
1	Vestas	Denmark	4,683	13.2%
2	Goldwind	China	3,654	10.3%
3	Enercon	Germany	3,583	10.1%
4	Siemens	Germany	2,838	8.0%
5	Suzlon Group	India	2,235	6.3%
6	GE	United States	1,738	4.9%
7	Gamesa	Spain	1,632	4.6%
8	United Power	China	1,384	3.9%
9	Mingyang	China	1,313	3.7%
10	Nordex	Germany	1,206	3.4%
11	XEMC	China	1,135	3.2%
12	Envision	China	1,100	3.1%
13	DEC	China	816	2.3%
14	Sinovel	China	816	2.3%
15	Sewind	China	780	2.2%
	Top-10		24,266	68.4%
	Other		6,563	18.5%
	TOTAL		**35,477**	

Source: Make, via Suzlon; Renewable Energy World, "Vestas Back on Top as Wind Turbine Installations Leader," March 12, 2014. http://www.renewableenergy world.com/rea/news/article/2014/03/vestas-back-on-top-as-wind-turbine-installions-leader.

Table A6. China's Grid Connectivity by Province, 2012
(Megawatts)

	Grid-connected capacity			Installed capacity			Percentage grid-connected
	MW	Share (%)	Rank	MW	Share (%)	Rank	Ratio
Inner Mongolia	14,384.0	30.1%	1	17,594.4	28.2%	1	81.8%
Gansu	5,551.6	11.6%	2	5,409.2	8.7%	3	102.6%*
Hebei	4,991.3	10.4%	3	6,969.5	11.2%	2	71.6%
Liaoning	4,039.5	8.4%	4	5,249.3	8.4%	4	77.0%
Jilin	2,936.3	6.1%	5	3,563.4	5.7%	6	82.4%
Shandong	2,718.6	5.7%	6	4,562.3	7.3%	5	59.6%
Heilongjiang	2,625.5	5.5%	7	3,445.8	5.5%	7	76.2%
Jiangsu	1,704.3	3.6%	8	2,886.2	4.6%	8	59.0%
Xinjiang	1,659.8	3.5%	9	2,316.1	3.7%	9	71.7%
Ningxia	1,361.5	2.8%	10	1,967.6	3.2%	10	69.2%
Shanxi	1,035.0	2.2%	11	1,881.1	3.0%	11	55.0%
Guangdong	933.0	2.0%	12	1,302.4	2.1%	12	71.6%
Fujian	873.7	1.8%	13	1,025.7	1.6%	13	85.2%
Yunnan	684.8	1.4%	14	932.3	1.5%	14	73.5%
Zhejiang	320.6	0.7%	15	367.2	0.6%	16	87.3%

Table A6. (Continued)

Shanghai	269.4	0.6%	16	318.0	0.5%	17	84.7%
Hainan	254.7	0.5%	17	256.7	0.4%	20	99.2%
Anhui	247.5	0.5%	18	297.0	0.5%	19	83.3%
Shaanxi	245.5	0.5%	19	497.5	0.8%	15	49.3%
Henan	154.0	0.3%	20	300.0	0.5%	18	51.3%
Beijing	150.0	0.3%	21	155.0	0.2%	24	96.8%
Hunan	133.8	0.3%	22	185.3	0.3%	23	72.2%
Jiangxi	133.5	0.3%	23	133.5	0.2%	25	100.0%
Tianjin	125.0	0.3%	24	243.5	0.4%	21	51.3%
Hubei	115.4	0.2%	25	100.4	0.2%	26	114.9%*
Guizhou	60.9	0.1%	26	195.1	0.3%	22	31.2%
Guangxi	49.5	0.1%	27	79.0	0.1%	27	62.7%
Chongqing	46.8	0.1%	28	46.8	0.1%	29	100.0%
Sichuan	16.0	0.0%	29	16.0	0.0%	30	100.0%
Qinghai	14.0	0.0%	30	67.5	0.1%	28	20.7%
Total	**47,835.5**	100.0%		**62,363.4**	100.0%		**77%**

Source: Li Junfeng et al., *China Wind Energy Outlook 2012* (Beijing: Chinese Renewable Energy Industries Association, Greenpeace, Global Wind Energy Council, and the Chinese Wind Energy Association, November 2012), pp. 6-7.

Table A7. U.S. System of Subsidies

	Amount '11-'15 budget ($ billions)	Term	Expiration	Type
Tax Credits				
Investment Tax Credit	$ 2.5	Upon installation	Year-end 2016	Foregone revenue
Production Tax Credit	$ 9.1	For 10 years	Year-end 2013	Foregone revenue
of which wind	$ *6.8*			
Advanced energy mfg tax credit (ARRA)	$ 1.4	Manufacturing	One-time (Jan10)	Foregone revenue
5-Year Cost Recovery for Renewable Energy Property	$ 1.1			
In lieu of tax credits				
1603 Grant (paid out in lieu of tax credits)	$ 17.2	Upon installation	Year-end 2011	Outlay
Total government spending (2011-2015)	**$ 31.3**			

Note: "ARRA" refers to American Recovery and Reinvestment Act.

Source: Congressional Research Service, *Renewable Energy Tax Incentives and Green Jobs*, June 19, 2012. *http://energycommerce.house.gov/sites/republicans.energycommerce.house.gov/files/Hearings/OI/20120619/HHRG-112-IF02-WStateMSherlock-20120619.pdf.*

Table A8. Turbine Installations in the United States by Manufacturer, 2005–2013 (Megawatts, share %)

		2005	2006	2007	2008	2009	2010	2011	2012	2013
	Turbine Installations (MW)									
1	GE Wind	1,431	1,146	2,342	3,585	3,995	2,543	2,006	5,014	984
2	Siemens	-	573	863	791	1,162	828	1,233	2,638	87
3	Vestas	699	439	948	1,120	1,489	221	1,969	1,818	4
4	Gamesa	50	74	494	616	600	566	154	1,341	-
5	Repower	-	-	-	94	330	68	172	595	-
6	Mitsubishi	190	128	356	516	814	350	320	420	-
7	Nordex	-	-	3	-	63	20	288	275	-
8	Clipper	3	-	48	470	605	70	258	250	-
9	Acciona	-	-	-	410	204	99	-	195	-
10	Suzlon	-	92	198	738	702	413	334	187	-
	Rest*	2	2	2	23	43	41	86	398	-
	Total	**2,374**	**2,453**	**5,253**	**8,362**	**10,005**	**5,220**	**6,819**	**13,131**	**1,087**
	Shares (%)									
	GE share	60.3%	46.7%	44.6%	42.9%	39.9%	48.7%	29.4%	38.2%	90.5%
	Other Top-10	39.7%	53.2%	55.4%	56.9%	59.7%	50.5%	69.3%	58.8%	8.4%
	Rest*	0.0%	0.0%	0.0%	0.3%	0.4%	0.8%	1.2%	3.0%	1.1%

*Rest in 2012 included 231 MW of Chinese manufacturer turbines.

Source: U.S. DOE, 2013 Wind Technologies Market Report (August 2014), p. 15. http://emp.lbl.gov/sites/all/files/2013_Wind_Technologies_Market_Report_Final3.pdf.

Table A9. Goldwind Financials, First Half of 2013 vs. First Half of 2014 (RMB millions)

	H12013	H12014
Financials (RMB millions)		
Pre-tax profit	109.6	376.8
Gross margin	18.6	30.4
Revenue from operations	3194.9	4400.0
Turbines and components	2830.1	3680.0
Wind power services	208.4	241.1
Power generation	186.3	520.3
Turnover		
Turbines sold (MW)	662.3	942.3
Turbines sold (units)	412.0	561.0
MW per unit	1.6	1.7

Source: Goldwind.

Table A10. Goldwind Revenue Breakdown, 2012–2013 (RMB millions; share %)

	RMB millions		Share (%)	
	2012	2013	2012	2013
Total revenues	11,225.00	12,196.00	100.0%	100.0%
By Market Segment				
Wind power services	*392.13*	*589.96*	3.5%	4.8%
Electric power generated at wind farms	*251.96*	*384.62*	2.2%	3.2%
Turbine sales	*10,580.91*	*11,221.42*	94.3%	92.0%
By Geography				
Mainland revenues	*9,920.00*	*10,834.61*	88.4%	88.8%
Overseas revenues	*1,305.00*	*1,361.39*	11.6%	11.2%

Source: *Hong Kong Government News*, "Xinjiang Goldwind Science & Technology: Goldwind Announces 2013 Annual Results, Distributed by Contify.com," March 21, 2014, via Factiva.

Table A11. Carbon Emissions and Carbon Intensity by Country

	CO2 emissions from fuel combustion						CO2 intensity at constant purchasing power parities					
	1990	2000	2006	2013	Change (%)		1990	2000	2006	2013	Change (%)	
	MtCO2				1990-2000	2000-2013	kCO2/$2005p				1990-2000	2000-2013
WORLD	20,490	22,787	27,140	31,097	11%	36%	0.564	0.472	0.452	0.415	-16%	-12%
China	**2,360**	**3,324**	**5,834**	**8,502**	**41%**	**156%**	**1.87**	**0.96**	**0.97**	**0.71**	**-49%**	**-26%**
United States	**4,827**	**5,601**	**5,618**	**5,101**	**16%**	**-9%**	**0.60**	**0.50**	**0.44**	**0.37**	**-17%**	**-26%**
India	548	939	1,257	2,011	71%	114%	0.52	0.51	0.45	0.46	-1%	-9%
Russia	2,341	1,501	1,575	1,661	-36%	11%	1.16	1.18	0.86	0.74	2%	-38%
Japan	1,005	1,134	1,149	1,186	13%	5%	0.32	0.31	0.29	0.29	-2%	-6%
Germany	961	810	796	764	-16%	-6%	0.46	0.33	0.30	0.27	-29%	-18%

Source: Enerdata.

Figures

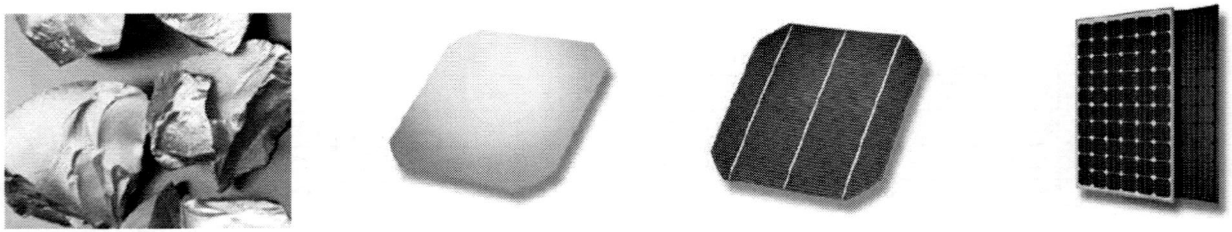

Sources: Vivian L, “China Solar: Who Survives China’s Polysilicon Shakeout?” Greentech Media, April 12, 2012; SolarWorld, “Solar Manufacturing Value Chain.” *http://www.solarworld-usa.com/about-solarworld/value-chain*.

Figure A1. Images of the Main Solar PV Components.

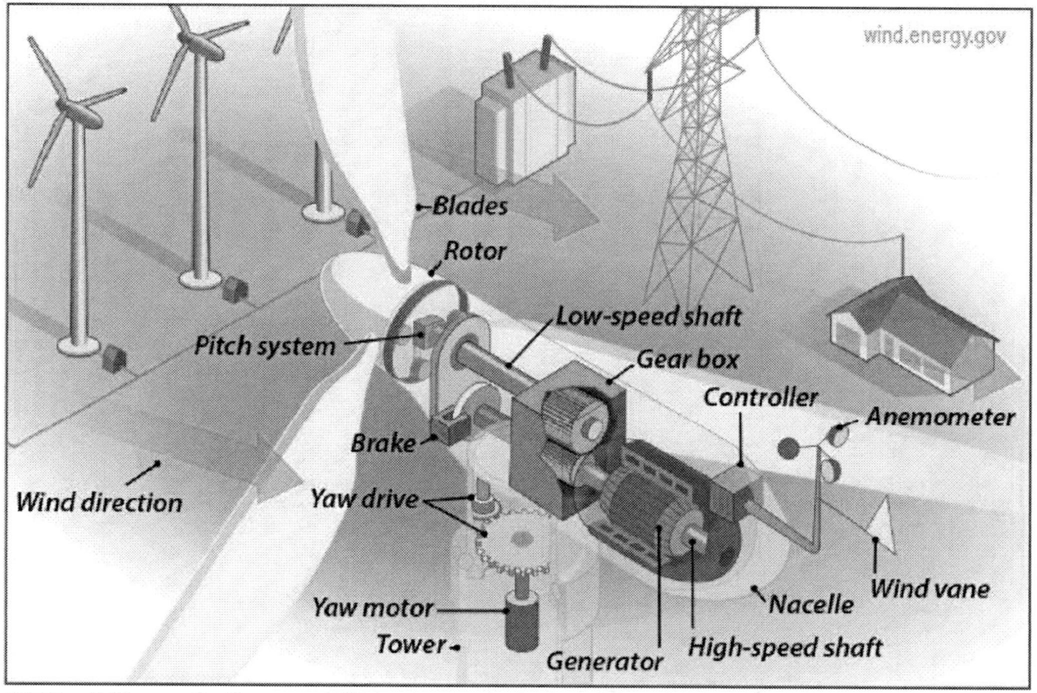

Source: U.S. DOE Office of Energy Efficiency and Renewable Energy, “The Inside of a Wind Turbine.” http://energy.gov/eere/wind/inside-wind-turbine-0.

Figure A2. Components of a Typical Wind Turbine.

REFERENCES

[1] For additional information, see the U.S.-China Economic and Security Review Commission (USCC), *2014 Annual Report to Congress*, November 2014, Chapter 1.4. See also the USCC's *Hearing on U.S.-China Clean Energy Cooperation: Status, Challenges, and Opportunities* (April 25, 2014), *Hearing on The Challenge of China's Green Technology Policy and Ohio's Response* (July 14, 2010); *Hearing on China's Green Energy and Environmental Policies* (April 8, 2010); and *Hearing on China's Energy Policies and Their Environmental Impacts* (August 13, 2008).

[2] For installed wind and solar capacity by country, see Appendix Tables A1 and A2.

[3] According to the U.S. Energy Information Administration: "Levelized cost of electricity (LCOE) is often cited as a convenient summary measure of the overall competiveness of different generating technologies. It represents the per kilowatt-hour cost (in real dollars) of building and operating a generating plant over an assumed financial life and duty cycle. Key inputs to calculating LCOE include capital costs, fuel costs, fixed and variable operations and maintenance (O&M) costs, financing costs, and an assumed utilization rate for each plant type. The importance of the factors varies among the technologies." U.S. Energy Information Administration, *Levelized Cost and Levelized Avoided Cost of New Generation Resources in the Annual Energy Outlook 2014*, April 17, 2014. *http://www.eia.gov/forecasts/aeo/electricity_generation.cfm.*

[4] In the 1950s, U.S. scientists at Bell Laboratories created the first commercial solar PV cell. The 1960s space race established the first wave of major U.S. solar PV manufacturers, though a small market and high prices hampered development. The oil crises of the 1970s brought forward the second generation of solar PV companies, many of which evolved into today's market leaders, such as First Solar. Michaela D. Platzer, *U.S. Solar Photovoltaic Manufacturing: Industry Trends, Global Competition, Federal Support* (Congressional Research Service, June 13, 2012), pp. 4-5.

[5] See Appendix Tables A4 and A5 for top-ten wind and solar manufacturers.

[6] There is some controversy with respect to these accounting methods and the definition of low-carbon goods and services. The UK government defines low-carbon environmental goods and services based on 24

sectors. These include not only renewable energy goods and services, such as solar, wind, and biomass, but also "environmental goods and services" (e.g., waste management) and "low carbon goods and services (e.g., building technologies, nuclear power, and carbon finance). UK Department for Business Innovation and Skills, *Low Carbon Environmental Goods and Services (LCEGS)*, July 2013. *https://www.gov.uk/government/publications/low-carbon-and-environmental-goods-and-services-2011-to-2012.*

[7] See Appendix Table A6 for grid connectivity by province in China in 2012.

[8] The net capacity factor of a power plant is the ratio of its actual output over a period of time, to its potential output if it were possible for it to operate at full nameplate capacity continuously over the same period of time.

[9] Baseload plants are the production facilities used to meet some or all of a given region's continuous energy demand, and produce energy at a constant rate, usually at a low cost relative to other production facilities available to the system.

[10] Equivalent data for the solar sector is difficult to obtain. The National Development and Reform Commission (NDRC), China's other regulatory agencies, and renewable energy industry associations do not publish figures on electricity generated by solar projects in China. There is a general consensus, however, that electricity generation from current projects is quite low, due to weak quality controls on installed equipment, curtailment, and grid infrastructure issues.

[11] See Appendix Table A7 for an overview of the principal U.S. subsidy programs.

[12] For critiques of China's industrial policy in the clean energy sector, see Rosalyn Hsue, *China's Regulatory State: A New Strategy for Globalization* (Ithaca: Cornell University Press, 2011), Chapter 8; Usha V. Haley, *Subsidies to Chinese Industry: State Capitalism, Business Strategy, and Trade Policy* (Oxford: Oxford University Press, 2013), Chapter 7.

[13] For more information, see the June 2013 edition of the *USCC Monthly Trade Bulletin*. The report can be accessed at *http://www.uscc.gov*.

[14] The EU also threatened China with antidumping duties on solar PV modules in 2013, which would have involved $27 billion worth of imports. Bilateral negotiations between the European Commission and China took place in summer 2013. The EU eventually agreed to a "price

undertaking," which essentially places a quota and a minimum price on panel imports from China but does not impose AD/CVD duties. European Commission, "EU Imposes Definitive Measures on Chinese Solar Panels, Confirms Undertaking with Chinese Solar Panel Exporters," December 2, 2013. *http://europa.eu/rapid/press-release_IP-13-1190_en.htm.*

[15] MOF covers 50 to 60 percent of the production costs of select solar companies, as well as 50 to 70 percent of installation costs for solar generation and distribution systems. Golden Sun projects were initially limited to a total capacity of 20 MW, but MOF soon revised this target upward, and China has since announced new rounds of subsidies. Becky Beetz and Wenjing Feng, "China: 2.83 GW of Solar to be Installed under Golden Sun Program," *PV Magazine*, December 11, 2012. *http://www.pv-magazine.com/news/details/beitrag/china--283- gw-of-solar-to-be-installed-under-golden-sun-program_100009542/#ixzz2Mao8mcXf.*

[16] According to the U.S. Solar Energy Industries Association, "Distributed generation (DG) refers to electricity that is produced at or near the point where it is used. Distributed solar energy can be located on rooftops or ground-mounted, and is typically connected to the local utility distribution grid." Solar Energy Industries Association, "Distributed Solar." *http://www.seia.org/policy/distributed-solar.*

[17] The United States has both general targets set by the DOE at the federal level, as well as state-level renewable portfolio standards. Data from the Database of State Incentives for Renewables & Efficiency indicates that as of 2013, 30 states and the District of Columbia had renewable portfolio standards (RPS), which require electricity providers to produce a certain percentage of energy from renewable sources. Six other states have set targets. Several states divide their RPS into different tiers, effectively creating more than one RPS target within their state. These targets vary considerably in terms of the target year and the target share for renewable energy mandated in each state. U.S. Energy Information Administration, "Energy in Brief: What Are Renewable Portfolio Standards (RPS) and How Do They Affect Renewable Electricity Generation?" January 25, 2013. *http://www.eia.gov/energy_in_brief/article/renewable_portfolio_standards.cfm.*

[18] Professor Keith makes the case for "policy-constrained wind power capacity." He argues that much existing research on the geophysical limits to global wind power rests on the assumption that one is willing to

cover all of the land or even all of the land and ocean surface with wind turbines. A more policy-relevant question is what amount of installed wind capacity is actually feasible. Wind power resources may look greater on paper, but could be limited by distance from energy demand, competition for land, and the quality of the wind resource. Existing studies operate under the rule of thumb that local power production could go as high as several watts per square meter. Professor Keith and his colleague find wind power extraction is actually below one watt per square meter. Amanda Adams and David Keith, "Are Global Wind Power Resource Estimates Overstated?" IOP Science *Environmental Research Letters* 8:1 (February 2013). *http://iopscience.iop.org/1748-9326/8/1/015021?v_showaffiliations=yes.*

[19] Shady Oaks is the first utility-scale wind power plant to use Goldwind's turbines in the Americas, with a 20-year power purchase agreement with Commonwealth Edison Co. Goldwind originally purchased the Shady Oaks wind project from Irish developer Mainstream Renewable Power. In December of 2012, Algonquin Power Co., Algonquin Power & Utilities Corp's non-regulated power generation subsidiary, acquired the project. Goldwind America, "Shady Oaks Wind Farm 109.5 MW." *http://www.goldwindamerica.com/project_profiles/shady-oaks-wind-farm/.*

[20] See Appendix Table A8 for the market share of individual companies in the U.S. wind turbine sector. The top Chinese turbine makers in the United States in 2012 were Goldwind (154.5 MW installed), followed at some distance by China Creative Wind Energy (61.2 MW), Guodian United Power (9 MW), Sinovel (4.5 MW), and Sany Electric (2 MW). *

[21] Sinovel's principal shareholders include Dalian Heavy Mechanical & Electrical (the Design Institute of Dalian Heavy Mechanical Group), where Sinovel's founder Wang Yuan previously worked; Beijing Tianhua Zhongtai, controlled by the founder, Han Junliang, one of China's richest men; New Horizon Capital, a China-focused private equity firm with ties to former Premier Wen Jiabao's son; and Tibet Xinmeng Investment Development, majority owned by Wei Wenyuan, formally a top executive at the Shanghai Stock Exchange. Denise Law, "Sinovel Wind's IPO to Raise Three Times More than Original Target," *Financial Times* Tilt, January 4, 2011. *http://tilt.ft.com/posts/2011-01/9711/sinovel-winds-ipo-to-raise-three-times-more-than-o;* Wayne Ma, "Sinovel Wind Group's Chairman Resigns, Citing Personal Reasons," *Wall Street Journal*, July 9, 2014. *http://www.wsj.com/articles*

/sinovel-wind-groups-chairmanresigns-citing-personal-reasons-1404870982; Russell Flannery, "China Mints Two Wind Power Billionaires as It Overtakes the U.S. in Capacity," Forbes, January 14, 2011. *http://www.forbes.com/sites/russellflannery/2011/01/14/china-mints-two-wind-power-billionairesas-it-overtakes-the-u-s-in-capacity/;* David Barboza, "Billions in Riches for Family of Chinese Leader," *New York Times*, October 25, 2012.

[22] For more information on Goldwind, see Appendix Tables A9 and A10 for company financials.

[23] Stock exchange data for Goldwind (002202:CH) and Sinovel (601558:CH) from Bloomberg.

[24] The China Securities Regulatory Commission has investigated Sinovel for alleged breaches of securities regulations. China's largest wind farm developer, Huaneng Renewables Corp. Ltd., has also sought damages against Sinovel in Chinese courts this year, on account of defective turbines and poor after-sales services. In October 2014, Huaneng Renewables sought $14.7 million in damages from Sinovel, based on allegations that Sinovel breached a 2013 supply deal by failing to resolve complaints about defective products. Beijing No.1 Intermediate People's Court heard the case on September 25. Huaneng claims it should not be required to pay the RMB 1.13 billion it owes Sinovel under their initial supply contract. Brian Publicover, "Huaneng Sues Sinovel for $14.7m," *Recharge News*, October 2, 2014. *http://www.rechargenews.com/wind/1378783/Huaneng-sues-Sinovel-for-14.7m;* Brian Publicover, "Sinovel Cut Q3 Losses as Orders Jump," *Recharge News*, October 31, 2014. *http://www.rechargenews.com/wind/1382162/Sinovel-cuts-Q3-loss-as-orders-jump.*

[25] It is difficult to gauge the revenues earned by U.S. companies from wind and solar technology licensed to China. U.S. government agencies do not report technology licensing revenues for the wind and solar sector specifically, in part because such data would require proprietary information from individual companies. A report by the National Foreign Trade Council does provide the source of technology for 70 wind turbine manufacturers in China. The list indicates that European designers—in particular the German companies Nordex, REpower, Dewind, and Aerodyn – are a more prevalent source of wind technology for Chinese companies than U.S. designers are. Dewey & LeBoeuf LLP (for the National Foreign Trade Council), *China's Promotion of the Renewable Electric Power Equipment Industry: Hydro, Wind, Solar,*

Biomass (Washington, DC: National Foreign Trade Council, March 2010), pp. 92-96.

[26] Data from U.S. International Trade Commission. HTS 8412909081.

[27] Data from the U.S. International Trade Commission. HTS 850231.

[28] For all definitions of wind and solar components, refer to Appendix Table A3 and Figures A1 and A2.

[29] For more information, see Woodrow Wilson Center China Environment Forum, "Electricity on the Move," February 19, 2013. *http://wilsoncenter.org/wilsonweekly/chinas-west-east-electricity-transfer-project.html.*

[30] The International Energy Agency defines energy security as "the uninterrupted availability of energy sources at an affordable price." International Energy Agency, "Energy Security." *http://www.iea.org /topics/energysecurity/.*

[31] For more information on the relocation of Chinese heavy industry to inland regions, see *USCC 2013 Annual Report*, November 2013, Chapter 1.1.

[32] Roughly 80 percent of China's oil travels through maritime chokepoints. To pursue alternatives, China is increasing pipeline-based imports; establishing a state-owned fleet of oil tankers; building up refinery capacity, and diversifying imports away from the Middle East and Africa. Overseas equity oil production entails financial and political risk, but offers China a means to offset oil price volatility. In addition, China began a strategic petroleum reserve (SPR) program in 2004, with the goal of having a reserve equivalent to 100 days of consumption by 2020. In recent months, declining oil prices have apparently allowed China to stock up its SPRs faster.

[33] China's indigenous coal supplies cover over two-thirds of the country's daily energy needs. However, China's coal use is unsustainable. The country has only 14 percent of the world's proven coal reserves but accounts for nearly half of global coal output. Coal is mined in the northwest of the country, from where it either needs to be shipped to power generation facilities in the demand centers along the east coast, or converted into power locally, to be transmitted eastward via high-voltage power lines. Both options are suboptimal: the former results in rail transport bottlenecks, while the latter wastes electricity in the process of transmission. Finally, there are environmental costs: coal treatment places a strain on China's scarce water supplies, and coal-fired generators are China's number-one source of air pollution.

[34] Hydraulic fracturing, or fracking, uses a high-pressure jet of sand, chemicals, and water to break up energy deposits buried deep underground to expose oil and gas resources. Drilling efficiency in the sector has improved to allow for more wells drilled at a faster pace than ever before, reducing the amount of rigs needed and yielding higher production rates. The media frequently links fracking with the shale gas revolution. However, fracking is also contributing to renewed oil production. Horizontal drilling, which drills lengthwise through the most energy-rich part of rock, is another new technique that has increased oil yields. Asjylyn Loder, "Fracking Pushes U.S. Oil Production to Highest Level in 20 Years," Bloomberg, January 9, 2013. *http://www.bloomberg.com/news/2013-01-09/frackingpushes-u-s-oil-production-to-highest-level-in-20-years.html.*

[35] An August 2013 Bloomberg report points to the risks of fracking in China's earthquake-prone Sichuan basin. Julio Friedmann, chief energy technologist at the Lawrence Livermore National Laboratory and former research scientist at Exxon Mobil Corp, stated: "The Sichuan basin is at the edge of the biggest continental collision in the world, India smashing into Asia. That's stressing the continental crust." A three-and-a-half year study by the Earthquake Administration Bureaux of Sichuan, Hebei and Zigong Municipality, recording more than 2,700 quakes of varying magnitude around an underground injection well in Zigong, Sichuan, found that "with the beginning of increased water-pressure injection, seismic activity around the test well showed a significant increase." Benjamin Haas, "China Fracking Quake-Prone Province Shows Zeal for Gas," Bloomberg, August 1, 2013. *http://www.bloomberg.com /news/articles/2013-07- 31/china-fracking-quake-prone-province-shows-zeal-for-gas.*

[36] The current debate over fracking in the United States may give pause to Chinese regulators. The U.S. Congress instructed the Environmental Protection Agency (EPA) in 2010 to determine the environmental impact of fracking. Originally due for completion in 2014, the EPA study has been postponed until 2016. See the EPA's preliminary study: U.S. Environmental Protection Agency, *Study of the Potential Impacts of Hydraulic Fracturing on Drinking Water Resources*, December 2012. *http://www.epa.gov/hfstudy/pdfs/hfreport20121214.pdf.*

[37] California was the first U.S. state to establish a cap-and-trade program, which went into effect in 2012 and introduced an enforceable compliance obligation for emissions cuts in 2013. By establishing a

price on carbon and reducing the carbon cap by 3 percent each year starting in 2013, California hopes to lower carbon emissions to 1990 levels by 2020 and achieve 80 percent reduction from 1990 levels by 2050. California Environmental Protection Agency Air Resources Board, "Cap-and-Trade Program," July 11, 2013. *http://www.arb.ca.gov /cc/capandtrade/capandtrade.htm.*

[38] For carbon intensity data, see Appendix Table A11.

[39] In June 2013, high-level officials from Shenzhen and California signed a memorandum of understanding for further cooperation on emissions trading systems. "Memorandum of Understanding to Enhance Cooperation on Emissions Trading Systems between the Government of Shenzhen Municipality of the People's Republic of China and the Government of the State of California of the United States of America," June 18, 2013. *http://www.arb.ca.gov/newsrel/2013/shenzhen_and_ ca_mou.pdf.*

End Notes

[1] The White House Office of the Press Secretary, United States and China Agree to Work Together on Phase Down of HFCs, June 8, 2013. http://www.whitehouse.gov/the-press-office/2013/06/08/united-states-and-china-agree-work-together-phase-down-hfcs.

[2] The White House Office of the Press Secretary, Fact Sheet: U.S.-China Joint Announcement on Climate Change and Clean Energy Cooperation, November 11, 2014. http://www.whitehouse.gov/the-press-office/2014/11/11/fact-sheet-us-china-joint-announcementclimate-change-and-clean-energy-c.

[3] Pew Charitable Trusts, 2013: Who's Winning the Clean Energy Race? (Washington, DC, April 2014), p. 12.

[4] U.S. Department of Energy, Revolution Now: The Future Arrives for Four Clean Energy Technologies – 2014 Update, October 2014, p. 2.

[5] Louise Downing, "U.S. Wind-Power Installations Rose Sixfold in 2014: BNEF," Bloomberg, January 22, 2015. http://www.bloomberg.com/news/articles/2015-01-22/u-s-wind-power-installations-rose-sixfold-in-2014-bnef.

[6] Pew Charitable Trusts, 2013: Who's Winning the Clean Energy Race? (Washington, DC, April 2014), pp. 7, 37, 50.

[7] Bloomberg New Energy Finance, "Press Release: Rebound in Clean Energy Investment in 2014 Beats Expectations," January 9, 2015.

[8] Zachary Shahan, "Solar PV Module Prices Have Fallen 80 Percent Since 2008, Wind Turbines 29 Percent," Clean Technica, May 6, 2013. http://cleantechnica.com/2013/05/06/solar-pv-module-prices-have-fallen-80-since-2008-wind-turbines-29/.

[9] U.S. Department of Energy, Revolution Now: The Future Arrives for Four Clean Energy Technologies – 2014 Update, October 2014, pp. 3, 5.

[10] U.S. Energy Information Administration, Levelized Cost and Levelized Avoided Cost of New Generation Resources in the Annual Energy Outlook 2014, April 17, 2014. http://www.eia.gov/forecasts/aeo/electricity_generation.cfm.

[11] Philip Andrews-Speed, Sufang Zhang, and Xiaoli Zhao, "Political and Institutional Analysis of the Successes and Failures of China's Wind Power Policy," Energy Policy 56 (2013): 333.

[12] U.S. Department of Energy, Revolution Now: The Future Arrives for Four Clean Energy Technologies – 2014 Update, October 2014, p. 3.

[13] Philip Andrews-Speed, Sufang Zhang, and Xiaoli Zhao, "Political and Institutional Analysis of the Successes and Failures of China's Wind Power Policy," Energy Policy 56 (2013): 333.

[14] U.S. Department of Energy, Revolution Now: The Future Arrives for Four Clean Energy Technologies – 2014 Update, October 2014, p. 6.

[15] Xinhua (English edition), "The World Can Relaunch Growth through Green Economy, Innovation," China.org.cn, September 11, 2009. http://www.china.org.cn/business /summerdavos2009/2009-09/11/content_18511815.htm.

[16] REN21, Renewables 2014 Global Status Report (Paris: REN21 Secretariat, 2014), p. 63. http://www.ren21.net/portals/0/documents/resources/gsr/2014/gsr2014_full%20reportiow% 20res.pdf.

[17] China National Bureau of Statistics, via CEIC; U.S. Energy Information Administration, "China." http://www.eia.gov/countries/analysisbriefs/China/china.pdf; William Gregozeski, "China Poised for Significant Expansion in Wind Power," Alt Energy Stocks, July 21, 2014.
http://www.altenergystocks.com/archives/2014/07/china_poised_for_significant_expansion _in_wind_power_generation.html; and John Romankiewicz (UC Berkeley), e-mail to Commission staff, January 8, 2015.

[18] Li Junfeng et al., China Wind Energy Outlook 2012 (Beijing: Chinese Renewable Energy Industries Association, Greenpeace, Global Wind Energy Council, and the Chinese Wind Energy Association, November 2012), p. 4.

[19] James Montgomery, "Wind Energy 2014 Outlook: Major Markets Recover, Battling Policy and Grid Concerns," Renewable Energy World Magazine, January/February 2014, pp. 34. http://www.gwec.net/wp-
content/uploads/2014/02/Wind_Energy_2014_Outlook_Sawyer.pdf.

[20] Doug Young, "China's Solar Energy Progress Sputters," Renewable Energy World, December 2, 2014. http://www.renewableenergyworld.com/rea/blog/post/2014/12/chinas-solar-energy-progress-sputters.

[21] Recharge News, "China Connected 18.7GW of Wind to the Grid in 2014, Says NEA," January 29, 2015. http://www.rechargenews.com/wind/1390102/china-connected-187gw-of-wind-to-the-grid-in-2014-says-nea.

[22] Michael Davidson, "Spilled Wind: An Update on China's Wind Integration Challenges," The Energy Collective, March 4, 2014. http://theenergycollective.com/michael-davidson/346951/spilled-wind-update-china-s-wind-integration-challenges.

[23] Energy Monitor Worldwide, "China Wind Sector Braces for FIT Cuts," September 19, 2014, via Factiva.

[24] Philip Andrews-Speed, Sufang Zhang, and Xiaoli Zhao, "Political and Institutional Analysis of the Successes and Failures of China's Wind Power Policy," Energy Policy 56 (2013): 337.

[25] Data from China National Renewable Energy Center. http://www.cnrec.org.cn/hd/2014-02-25-412.html; Lori Bird, Jaquelin Cochran, and Xi Wang, Wind and Solar Energy Curtailment: Experience and Practices in the United States. Technical Report NREL/TP-6A20- 60983.

(Denver, CO: National Renewable Energy Laboratory), p. 3. http://www.nrel.gov/docs/fy14osti/60983.pdf.

[26] Liu Yuanyuan, "China Curtails Less Wind Energy in 2014," Renewable Energy World, November 20, 2014. http://www.renewableenergyworld.com/rea/news/article/2014/11/china-curtails-less-wind-energy-in-2014.

[27] House Oversight and Government Reform Subcommittee on Energy Policy, Hearing on Oversight of the Production Tax Credit, written testimony of Rob Gramlich, October 2, 2013.

[28] Hugh Shaman, Henrik Meyer, and Martin Agerup, Wind Energy: The Case of Denmark (Copenhagen, Denmark: Center for Politiske Studies, September 2009).

[29] Data from U.S. Energy Information Administration.

[30] John Romanikiewicz, "Crystal Ball: China Will Not Produce More Wind Energy than the U.S. before 2020," Green Leap Forward (Blog), February 26, 2013. http://greenleapforward.com/2013/02/26/crystal-ball-china-will-not-produce-more-wind-energy-than-the-us-before-2020/.

[31] Economist, "Blown Away," June 8, 2013. http://www.economist.com/news/united-states/21579046-wind-power-doing-well-it-still-reliesirregular-and-short-term-subsidies-blown.

[32] Pew Charitable Trusts, 2012: Who's Winning the Clean Energy Race? (Washington, DC: April 2013), p. 18.

[33] Joseph E. Aldy, A Preliminary Review of the American Recovery and Reinvestment Act's Clean Energy Package, Faculty Research Working Paper Series RWP11-048 (Cambridge, MA: Harvard Kennedy School, December 2011).

[34] House Oversight and Government Reform Subcommittee on Energy Policy, Hearing on Oversight of the Production Tax Credit, written testimony of Rob Gramlich, October 2, 2013.

[35] Melanie Hart (Director, China Policy, Center for American Progress), e-mail to Commission staff, January 9, 2015.

[36] House Oversight and Government Reform Subcommittee on Energy Policy, Hearing on Oversight of the Production Tax Credit, written testimony of Robert J. Michaels, October 2, 2013.

[37] Bloomberg New Energy Finance, "Press Release: Rebound in Clean Energy Investment in 2014 Beats Expectations," January 9, 2015.

[38] London Array, "London Array Fully Powered and Producing Green Electricity." April 8, 2013. http://www.londonarray.com/2013/04/11/london-array-fully-powered-and-producing-green-electricity/.

[39] The European Wind Energy Association, Wind in Our Sails: The Coming of Europe's Offshore Wind Energy Industry (November 2011), pp. 8, 11. http://www.ewea.org/fileadmin/ewea_documents/documents/publications/reports/23420_Offshore_report_web.pdf.

[40] U.S. Department of Energy and U.S. Department of the Interior, A National Offshore Wind Industry: Creating an Offshore Wind Energy Industry in the United States (Washington, DC, February 2011).

[41] Li Junfeng et al., China Wind Energy Outlook 2012 (Beijing: Chinese Renewable Energy Industries Association, Greenpeace, Global Wind Energy Council and the Chinese Wind Energy Association, November 2012), pp. 15-16; Manuela Zoninsein, "Chinese Offshore Development Blows Past U.S.," New York Times, September 7, 2010. http://www.nytimes.com/cwire/2010/09/07/07climatewire-chineseoffshore-development-blows-past-us-47150.html.

[42] Bloomberg, "China Three Years Late on Installing Offshore Wind Farms," July 16, 2014. http://www.bloomberg.com/news/2014-07-16/china-three-years-late-on-installing-offshore-wind-farms.html.

[43] Bloomberg, "China Three Years Late in Installing Offshore Wind Farms," July 16, 2014. http://www.bloomberg.com/news/2014-07-16/china-three-years-late-on-installing-offshore-wind-farms.html.

[44] U.S. Department of Energy, Maine Project Launches First Grid-Connected Offshore Wind Turbine in the U.S., May 31, 2013. http://energy.gov/articles/maine-project-launches-first-grid-connected-offshore-wind-turbine-us.

[45] Cape Wind, "Cape Wind Project Status and Timeline." http://www.capewind.org/when.

[46] U.S. Energy Information Administration, Levelized Cost and Levelized Avoided Cost of New Generation Resources in the Annual Energy Outlook 2014, April 17, 2014. http://www.eia.gov/forecasts/aeo/electricity_generation.cfm.

[47] Liu Yuanyuan, "Photovoltaic Production in China's Jiangsu Province Falls 30 Percent in 2012," Renewable Energy World, February 13, 2013. http://www.renewableenergyworld.com/rea/news/print/article/2013/02/photovoltaic-production-in-chinas-jiangsu-province-falls30-in-2012.

[48] U.S. Department of Energy, SunShot Initiative, February 7, 2013. http://www1.eere.energy.gov/solar/sunshot/about.html.

[49] United States International Trade Commission (HTS code 8541406020).

[50] International Trade Commission, Hearing on Crystalline Silicon Photovoltaic Cells and Modules from the People's Republic of China, Antidumping and Countervailing Duty Investigation, written testimony of Joseph Morinville, October 3, 2012; U.S.-China Economic and Security Review Commission, Hearing on The Challenge of China's Green Technology Policy and Ohio's Response, written testimony of David McCall, July 14, 2010.

[51] Inside US-China Trade, "Experts Debate Impact of Chinese Policies on Clean Energy Innovation," November 22, 2011.

[52] Nathaniel Ahrens, China's Competitiveness: Myth, Reality, and Lessons for the United States – Case Study: Suntech (Washington, DC: Center for Strategic and International Studies, January 2013), p. 21.

[53] Dewey & LeBoeuf LLP (for the National Foreign Trade Council), China's Promotion of the Renewable Electric Power Equipment Industry: Hydro, Wind, Solar, Biomass (Washington, DC: National Foreign Trade Council, March 2010), pp. 23-24.

[54] Dewey & LeBoeuf LLP (for the National Foreign Trade Council), China's Promotion of the Renewable Electric Power Equipment Industry: Hydro, Wind, Solar, Biomass (Washington, DC: National Foreign Trade Council, March 2010).

[55] Cesar Tordesillas, "China Signs 2.3 GW Project Agreements in Pakistan," Asian Power, December 21, 2010. http://asianpower.com/project/in-focus/china-signs-23-gw-project-agreements-in-pakistan; Rashid Ahman Khan, "The Pakistan-China Strategic Partnership," China.org.cn, May 20, 2011. http://www.china.org.cn/opinion/2011-05/20/content_22605398_2.htm.

[56] Sally Bakewell, "Goldwind Signs $5.5 Billion China Development Bank Wind Pact," Bloomberg, January 31, 2012. http://www.bloomberg.com/news/2012-01-31/goldwind-signs-5-5-billion-china-development-bank-pact-for-wind.html.

[57] Bloomberg, "China Development Bank Funds Wind Projects at Guangdong Nuclear," November 22, 2012. http://www.businessweek.com/news/2012-11-22/china-development-bank-funds-wind-projects-at-guangdong-nuclear.

[58] Jianxiang Yang, "China Development Bank Funds Sinovel Project in Brazil," Wind Power Monthly, September 11, 2012. http://www.windpowermonthly.com/article/1149254/china-development-bank-funds-sinovel-project-brazil.

[59] Inside US-China Trade, "Experts Debate Impact of Chinese Policies on Clean Energy Innovation," November 22, 2011.

[60] Office of the United States Trade Representative, United States Details China and India Subsidy Programs in Submission to WTO, October 2011. http://www.ustr.gov/about-us/press-office/press-releases/2011/october/united-states-details-china-and-india-subsidy-prog.

[61] Charles W. Thurston, "ITC Makes Unanimous Decision on Chinese Cell & Modules," Renewable Energy World, November 7, 2012. http://www.renewableenergyworld.com /rea/news/article/2012/11/commerce-slams-chinese-cell-modules-with-final-duties.

[62] Inside U.S. Trade, "DOC Levels High Final Duties on Chinese and Taiwanese Solar Products," December 19, 2014. http://chinatradeextra.com/Inside-US-Trade/Inside-U.S.-Trade-12/19/2014/doc-levels-high-final-duties-on-chinese-and-taiwanesesolar-products/menu-id-710.html.

[63] Inside U.S.-China Trade, "Commerce Says It May Slash AD Rates under First China Solar Case, Following Review," January 7, 2015. http://chinatradeextra.com/201501072486566 /China-Trade-Extra-General/Daily-News/commerce-says-it-may-slash-ad-rates-underfirst-china-solar-case-following-review/menu-id-428.html.

[64] Inside U.S. Trade, "DOC Levels High Final Duties on Chinese and Taiwanese Solar Products," December 19, 2014. http://chinatradeextra.com/Inside-US-Trade/Inside-U.S.-Trade-12/19/2014/doc-levels-high-final-duties-on-chinese-and-taiwanesesolar-products/menu-id-710.html.

[65] Inside U.S. Trade, "Commerce Says It May Slash AD Rates under First China Solar Case, Following Review," January 7, 2015. http://chinatradeextra.com/201501072486566/China-Trade-Extra-General/Daily-News/commerce-says-it-may-slash-ad-rates-underfirst-china-solar-case-following-review/menu-id-428.html.

[66] Michael Levi, "Tilting at Wind Turbines," Foreign Policy, January 19, 2011.

[67] Diane Cardwell, "China's Feud with West on Solar Leads to Tax," New York Times, July 18, 2013. http://www.nytimes.com/2013/07/19/business/energy-environment/chinas-feud-with-west-on-solar-leads-to-tax.html.

[68] Melanie Hart (Director, China Policy, Center for American Progress), e-mail to Commission staff, January 9, 2015.

[69] Wang Qiang, "Effective Policies for Renewable Energy—The Example of China's Wind Power—Lessons for China's Photovoltaic Power," Renewable and Sustainable Reviews (2010): 14, 704-705.

[70] National Development and Reform Commission, The Renewable Energy Law of the People's Republic of China, February 28, 2005. Staff translation.

[71] Joanna Lewis, Zhongying Wang, and Haiyan Qin, "China's Wind Power Industry: Policy Support, Technological Achievements, and Emerging Challenges," Energy Policy 51 (2012): 81-82.

[72] National Development and Reform Commission, Medium- and Long-Term Development Plan for Renewable Energy in China, September 2007. Staff translation.

[73] Li Junfeng et al., China Wind Energy Outlook 2012 (Beijing: Chinese Renewable Energy Industries Association, Greenpeace, Global Wind Energy Council, and the Chinese Wind Energy Association, November 2012), p. 4.

[74] World Resources Institute, "Why Is China Taking Action on Clean Energy and Climate Change?" ChinaFAQs Issue Brief (May 2013), p. 4.

[75] Bloomberg New Energy Finance, Joined at the Hip: The U.S.-China Clean Energy Relationship. White Paper. (May 2010), p. 8.

[76] REN21, Renewables 2014 Global Status Report (REN21 Secretariat, June 6, 2014), p. 111.

[77] Julian Wong, "Dawn of a New Era: The Gansu Solar Concession and Landmark Solar Roofs Program," The Green Leap Forward, March 27, 2009. http://greenleapforward.com/2009/03/27/dawn-of-a-new-era-the-gansu-solar-concession-and-landmark-solar-roofsprogram/.

[78] Richard J. Campbell, China and the United States – A Comparison of Green Energy Programs and Policies (Congressional Research Service, June 14, 2010), pp. 8-9; Coco Liu, "China Uses Feed-In Tariff to Build Domestic Solar Market," New York Times, September 14, 2011. http://www.nytimes.com/cwire/2011/09/14/14climatewire-china-uses-feed-in-tariff-to-build-domestic-25559.html.

[79] Xinhua (English edition), "New Policies Support Ailing Solar Industry," October 29, 2012. http://usa.chinadaily.com.cn/business/2012- 10/29/content_15853900.htm.

[80] Andy Colthorpe, "China to Target 35GW of Solar by 2015," PV Tech, July 15, 2013. http://www.pvtech.org/news/china_to_target_35gw_of_solar_by_2015.

[81] Bloomberg, "China Rushes to Harness Wind While Government Pays," October 29, 2014. http://www.bloomberg.com/news/2014-10- 29/china-rushes-to-harness-wind-while-government-still-pays.html; Louise Downing, "U.S. Wind-Power Installations Rose Sixfold in 2014: BNEF," Bloomberg, January 22, 2015. http://www.bloomberg.com/news/articles/2015-01-22/u-s-wind-power-installations-rosesixfold-in-2014-bnef.

[82] Xinhua (English edition), "China Starts Building 'Three Gorges in the Air,'" August 8, 2009. http://www.chinadaily.com.cn/china/2009- 08/08/content_8546066.htm.

[83] Li Junfeng et al., China Wind Energy Outlook 2012 (Beijing: Chinese Renewable Energy Industries Association, Greenpeace, Global Wind Energy Council, and the Chinese Wind Energy Association, November 2012), pp. 9-10.

[84] Eric Ng, "Wind Power Firms Fall as Tariff Cuts Loom," South China Morning Post, September 16, 2014, via Factiva.

[85] Bloomberg, "China Rushes to Harness Wind While Government Pays," October 29, 2014; Energy Monitor Worldwide, "China Wind Sector Braces for FIT Cuts," September 19, 2014, via Factiva. http://www.bloomberg.com/news/2014-10-29/china-rushes-to-harness-wind-while-government-still-pays.html.

[86] Jianxiang Yang, "China Confirms Cut to Onshore Wind Tariff," Wind Power Monthly, January 8, 2015. http://www.windpowermonthly.com/article/1328437/china-confirms-cut-onshore-tariff.

[87] Recharge News, "China Connected 18.7GW of Wind to the Grid in 2014, Says NEA," January 29, 2015. http://www.rechargenews.com/wind/1390102/china-connected-187gw-of-wind-to-the-grid-in-2014-says-nea.

[88] Brian Publicover, "China 'to End Curtailment' in 2015," Recharge News, October 22, 2014. http://www.rechargenews.com/wind/1380913/China-to-end-curtailment-in-2015.

[89] Ma Chunbo and He Lining, "From State Monopoly to Renewable Portfolio: Restructuring China's Electric Utility," Energy Policy 36 (2008): 1703-1705; Todd J. Edwards, "China's Power Sector Restructuring and Electricity Price Reforms," Asia Papers 6(2) (Brussels, Belgium: Brussels Institute of Contemporary China Studies, January 18, 2012), pp. 17-20.

[90] Joanna Lewis, Zhongying Wang and Haiyan Qin, "China's Wind Power Industry: Policy Support, Technological Achievements, and Emerging Challenges," Energy Policy 51 (2012): 86.

[91] Judith A. Cherni and Joanna Kentish, "Renewable Energy Policy and Electricity Market Reforms in China," Energy Policy 35 (2007): 3623-3624.

[92] Todd J. Edwards, "China's Power Sector Restructuring and Electricity Price Reforms," Asia Papers 6(2) (Brussels, Belgium: Brussels Institute of Contemporary China Studies, January 18, 2012), pp. 19-22.

[93] World Energy Council, 2013 Energy Sustainability Index (London, UK, 2013). http://www.worldenergy.org/wpcontent/uploads/2013/09/2013-Energy-Sustainability-Index-VOL-2.pdf.

[94] CCTV, "China Launched New Electricity Pricing Tariff," July 1, 2012. http://english.cntv.cn/program/china24/20120701/107372.shtml.

[95] Ryan Rutkowski, Rebalancing and Rising Electricity Prices in China. China Economic Watch (Peterson Institute for International Economics, February 18, 2014). http://blogs.piie.com/china/?p=3760.

[96] Axel Berger, Rasmus Lema, and Hubert Schmitz, "China's Impact on the Global Wind Power Industry," Journal of Current Chinese Affairs 42 no. 1 (2013): 47.

[97] Li Junfeng et al., China Wind Energy Outlook 2012 (Beijing: Chinese Renewable Energy Industries Association, Greenpeace, Global Wind Energy Council, and the Chinese Wind Energy Association, November 2012), pp. 12-13 and Executive Summary; Axel Berger, Rasmus Lema, and Hubert Schmitz, "China's Impact on the Global Wind Power Industry," Journal of Current Chinese Affairs 42 no. 1 (2013): 47; Philip Andrews-Speed, Sufang Zhang, and Xiaoli Zhao, "Political and Institutional Analysis of the Successes and Failures of China's Wind Power Policy," Energy Policy 56 (2013): 338.

[98] Priya Barua, Letha Tawney and Lutz Weischer, Delivering on the Clean Energy Economy: The Role of Policy in Developing Successful Solar and Wind Industries (World Resources Institute, September 2012), p. 35.

[99] Li Shuo, "People Power: Unlocking China's Domestic Solar Market," Greenpeace Blog, August 1, 2013. http://www.greenpeace.org/eastasia/news/blog/people-power-unlocking-chinas-domestic-solar-/blog/46018/.

[100] Solar and Wind Energy Resource Assessment (SWERA), "Wind Maps." http://en.openei.org/wiki/SWERA/Data.

[101] Lewis, Joanna, Zhongying Wang, and Haiyan Qin, "China's Wind Power Industry: Policy Support, Technological Achievements, and Emerging Challenges," Energy Policy 51 (2012): 87.

[102] Eric Ng, "China to See Lower Growth in Solar Farm Capacity This Year," South China Morning Post, November 24, 2014. http://www.scmp.com/business/china-business/article/1647172/china-see-lower-growth-solar-farm-capacity-year.

[103] U.S. Energy Information Administration, China Analysis Brief, April 22, 2013. http://www.eia.gov/countries/analysisbriefs/China/china.pdf; White House Office of the Press Secretary, Fact Sheet: U.S.-China Joint Announcement on Climate Change and Clean Energy Cooperation, November 11, 2014. http://www.whitehouse.gov/the-press-office/2014/11/11/fact-sheet-us-china-joint-announcement-climate-change-and-clean-energy-c.

[104] Xiaomei Tan et al., China's Overseas Investments in the Wind and Solar Industries (Washington, DC: World Resources Institute, April 2013).

[105] Keith Bradsher, "Suntech Unit Declares Bankruptcy," New York Times, March 20, 2013. http://www.nytimes.com/2013/03/21/business/energy-environment/suntech-declares-bankruptcy-china-says.html.

[106] Lucy Hornby, "Shunfeng Scoops up Bankrupt Suntech Unit," Financial Times, November 3, 2013. http://www.ft.com/cms/s/0/0ca3fc98- 4455-11e3-a751-00144feabdc0.html#ixzz3GFaxghE3.

[107] Bloomberg, "China's Wind Turbine Makers Face Market Consolidation," April 17, 2014. http://www.bloomberg.com/news/2014-04-17/china-s-wind-turbine-makers-face-consolidation-as-glut-lingers.html.

[108] Goldwind, "Wind Turbine Sales." http://www.goldwindamerica.com/technology-capabilities/wind-turbine-sales/; CQ FD Disclosure, "Q4 2013 China Ming Yang Wind Power Group Limited Earnings Conference Call – Final," April 7, 2014, via Factiva.

[109] Hong Kong Government News, "Xinjiang Goldwind Science & Technology: Goldwind Announces 2013 Annual Results," March 21, 2014, via Factiva; SeeNews Renewables, "Goldwind's 2013 Net Profit Almost Triples," March 24, 2014, via Factiva; SinoCast Energy Beat, "Goldwind 2013 Net Up 179 percent," March 24, 2014, via Factiva.

[110] Goldwind America, "Shady Oaks Wind Farm 109.5 MW." http://www.goldwindamerica.com/project_profiles/shady-oaks-wind-farm/.

[111] Mena Report, "To Supply 51 MW Wind Turbine. Under the Contract, 34 Units of Goldwind's 1.5 MW Direct-Drive Permanent Magnet (DDPM) Wind Turbines Will Be Delivered to Cuba," December 20, 2013, via Factiva.

[112] ET Net News, "Goldwind's First 2.5 MW Wind Turbines Installed in Romania," December 5, 2013, via Factiva.

[113] SeeNews Renewables, "Goldwind USA Grabs 215-MW Turbine Order in Panama," April 25, 2014, via Factiva.

[114] Stephen Bierman, "E4 Group Says Goldwind Seeks to Build Wind Blade Plant in Russia," Bloomberg, August 28, 2014. http://www.bloomberg.com/news/2014-08-28/e4-group-says-goldwind-seeks-to-build-wind-blade-plant-in-russia.html.

[115] Wind Power Monthly, "Goldwind to Build Factory in Portugal," December 1, 2011. http://www.windpowermonthly.com/article/1107399/goldwind-build-factory-portugal.

[116] CQ FD Disclosure, "Q4 2013 China Ming Yang Wind Power Group Limited Earnings Conference Call – Final," April 7, 2014, via Factiva.

[117] Hoovers, "Sinovel Wind Group Co., Ltd. Profile." December 19, 2011.Dewey & LeBoeuf LLP (for the National Foreign Trade Council); China's Promotion of the Renewable Electric Power Equipment Industry: Hydro, Wind, Solar, Biomass (Washington, DC: National Foreign Trade Council, March 2010), p. 92.

[118] Sally Bakewell, "Goldwind Signs $5.5 Billion China Development Bank Wind Pact," Bloomberg, January 31, 2012. http://www.bloomberg.com/news/2012-01-31/goldwind-signs-5-5-billion-china-development-bank-pact-for-wind.html.

[119] See also Goldwind, "History." http://www.goldwindamerica.com/about-goldwind/history/; Iacob Koch-Weser and Regina Abrami, Goldwind USA: Chinese Wind in the Americas (Cambridge, MA: Harvard Business School, May 3, 2012).

[120] Bloomberg, "Goldwind Says Annual Profit May Quadruple on Higher Demand," October 29, 2014. http://www.bloomberg.com/news/2014-10-29/goldwind-says-annual-profit-may-quadruple-on-higher-demand.html.

[121] David Yong, "Sinovel Bonds Tumble after Trading Halt Caution as Losses Widen," Bloomberg, March 27, 2014. http://www.bloomberg.com/news/2014-03-27/sinovel-bonds-tumble-after-trading-halt-caution-as-losses-widen.html.

[122] Jianxiang Yang, "Shanghai Exchange Confirms Sinovel Suspension," Wind Power Offshore, May 14, 2014. http://www.windpoweroffshore.com/article/1294407/shanghai-exchange-confirms-sinovel-suspension.

[123] Patrick Smith and Jianxiang Yang, "Sinovel Warns of 'Unpredictable' Impact of AMSC Litigation," Wind Power Monthly, September 1, 2014. http://www.windpowermonthly.com /article/1310118/sinovel-warns-unpredictable-impact-amsc-litigation.

[124] Joan Lappin, "American Superconductor and Its Rogue Employee Both Duped by Sinovel," Forbes, September 27, 2011. http://www.forbes.com/sites/joanlappin/2011/09/27/american-superconductor-and-its-rogue-employee-both-duped-by-sinovel/.

[125] Ed Crooks, "US Charges Sinovel with Trade Secret Theft," Financial Times, June 28, 2013. http://www.ft.com/intl/cms/s/0/796f78d2-df67-11e2-a9f4-00144feab7de.html#axzz3Jedpnzop.

[126] Patrick Smith and Jianxiang Yang, "Sinovel Warns of 'Unpredictable' Impact of AMSC Litigation," Wind Power Monthly, September 1, 2014. http://www.windpowermonthly.com /article/1310118/sinovel-warns-unpredictable-impact-amsc-litigation.

[127] Konrad J. A. Kundig, "Copper-Based Solar Cells: Good for the Environment, Good for the Consumer," Copper.org, May 2007. http://www.copper /innovations/2007/05/solar_energy.html.

[128] Richard J. Campbell, China and the United States – A Comparison of Green Energy Programs and Policies (Washington, DC: Congressional Research Service, June 2010), p. 8.

[129] Shyam Mehta, "Polysilicon 2015-2018: Supply, Demand, Cost, and Pricing," GTM Research, October 14, 2014. http://www.greentechmedia.com/research/report/polysilicon-2015-2018.

[130] Mark Osborne, "China Report: Six Polysilicon Producers Left Standing," PVTech, July 10, 2013. http://www.pvtech.org/news/china_report_six_polysilicon_producers_left_standing.

[131] U.S. Department of Energy, 2011 Critical Materials Strategy (Washington, DC, December 2011), p. 20. http://energy.gov/sites/prod/files/DOE_CMS2011_FINAL_Full.pdf.

[132] Inside U.S. Trade China Trade Extra, "Appellate Body Upholds Panel Findings in Rare Earths Case, Handing U.S. Victory," August 8, 2014.

[133] Marc Humphries, Rare Earth Elements: The Global Supply Chain (Washington, DC: Congressional Research Service, December 16, 2013), p. 19. https://www.fas.org/sgp /crs/natsec/R41347.pdf.

[134] Inside U.S. Trade China Trade Extra, "Appellate Body Upholds Panel Findings in Rare Earths Case, Handing U.S. Victory," August 8, 2014.

[135] Chuin-Wei Yap, "China Ends Rare-Earth Minerals Export Quotas," Wall Street Journal, January 5, 2015. http://www.wsj.com/articles/china-ends-rare-earth-minerals-export-quotas-1420441285; Hongpo Shen, "China Ends 'Export Quota' System and Counters with a 'Strict Export License' to Limit the World's Supply of Rare Earths, Tungsten and Molybdenum," InvestorIntel, January 4, 2015. http://investorintel.com/rare-earth-intel/china-ends-export-quota-system-counters-strict-exportlicensepolicy-limit-worlds-supply-rare-earths-tungsten-molybdenum/.

[136] John Kemp, "Column – Super Grid: China Masters Long-Distance Power Transmission," Reuters, June 19, 2014. http://www.reuters.com/article/2014/06/19/us-china-electricity-grid-kemp-idUSKBN0EU19B20140619.

[137] Peter Kelly-Detwiler, "Low Carbon Economies and Smart Grids: Watch Out For China," Forbes, March 26, 2013. http://www.forbes.com/sites/peterdetwiler/2013/03/26/low-carbon-economies-and-smart-grids-watch-out-for-china/.

[138] Bloomberg New Energy Finance, "China Out-Spends the US for First Time in $15bn Smart Grid Market," February 18, 2014. http://about.bnef.com/press-releases/china-out-spends-the-us-for-first-time-in-15bn-smart-grid-market/.

[139] Andy Colthorpe, "IHS: Grid-connected Energy Storage to Exceed 40GW by 2022," PV Tech, January 16, 2014. http://www.pvtech.org/news/ihs_grid_connected_energy exceed_40gw_by_2022.

[140] California Public Utilities Commission, "CPUC Sets Energy Storage Goals for Utilities," October 17, 2013. http://docs.cpuc.ca.gov/PublishedDocs/Published/G000 /M079/K171 /79171502.PDF.

[141] U.S. Department of Energy, 2011 Critical Materials Strategy (Washington, DC, December 2011), p. 20.

[142] U.S. Department of Energy, 2013 Wind Technologies Market Report (Washington, DC, 2014), pp. 30-31. http://emp.lbl.gov/sites/all/files/2013_Wind_Technologies_Market_ Report_Final3.pdf.

[143] U.S. Department of Energy, 2012 Wind Technologies Market Report (Washington, DC, 2013), p. 14. http://emp.lbl.gov/sites/all/files/lbnl-6356e.pdf.

[144] Goldwind, "Goldwind Completes First Large-Scale High Altitude Wind Project," December 29, 2011. http://www.goldwindglobal.com/web/news.do?action=detail&id=201112290303128872.

[145] First Solar, "First Solar Sets New World Record For CDTE Solar Cell Efficiency," February 26, 2013. http://investor.firstsolar.com/releasedetail.cfm?releaseid=743398; Renewable Energy World, "First Solar Reaches US $1 Per Watt Milestone," February 25, 2009. http://www.renewableenergyworld.com/rea/news/article/2009/02/first-solar-passes-us1watt.

[146] General Electric, "GE Developing Wind Blades That Could Be the "Fabric" of Our Clean Energy Future," November 28, 2012. http://www.geglobalresearch.com/news/press-releases/ge-developing-wind-blades-that-could-be-the-fabric-of-our-clean-energy-future.

[147] National Wind Watch, "FAQ – Size." https://www.wind-watch.org/faq-size.php; U.S. Department of Energy, "Advanced Blade Manufacturing." http://energy /advanced-blade-manufacturing.

[148] Michaela D. Platzer, U.S. Wind Turbine Manufacturing: Federal Support for an Emerging Industry (Washington, DC: Congressional Research Service, December 2012), p. 9.

[149] Virginia Tech, "CREATe – Renewable Energy – Wind Turbine Aerodynamics." http://www.create.centers.vt.edu/research/renewableenergy.html; Michaela D. Platzer, U.S. Wind Turbine Manufacturing: Federal Support for an Emerging Industry (Washington, DC, December 18, 2012), pp. 10-14.

[150] Goldwind, "Why PMDD?" http://www.goldwindamerica.com/technology-capabilities/pmdd/.

[151] Thilo Grau, Molin Huo, and Karsten Neuhoff, "Survey of Photovoltaic Industry and Policy in Germany and China," Energy Policy 51 (2012): 22.

[152] Jonathan Gifford, "Soitec Achieves 46% Multi-Junction Cell for CPV," PV Magazine, December 2, 2014. http://www.pvmagazine.com/news/details/beitrag/soitec-achieves-46-multi-junction-cell-for-cpv_100017342/#axzz3KmmUOXld.

[153] Sarah Kurtz, "High-Efficiency, Multi-Junction Solar Cells for Large-Scale Solar Electricity Generation." Presentation by the National Renewable Energy Laboratory at APS March Meeting, 2006. http://www.aps.org/meetings/multimedia Multijunction_ Solar_Cells_for_Large_Scale_Solar_Electricity_Gener ation_Kurtz.pdf.

[154] Mark Shwartz, "Stanford Scientists Build the First All-Carbon Solar Cell," Stanford Report, October 31, 2012. http://news.stanford.edu/news/2012/october/carbon-solar-cell-103112.html.

[155] Eric Gershon, "New Carbon Films Improve Prospects of Solar Energy Devices," Yale University News, February 13, 2013. http://news.yale.edu/2013/02/13/new-carbon-films-improve-prospects-solar-energy-devices.

[156] Darren Quick, "First True 'All-Carbon' Solar Cell Developed," Gizmag, October 31, 2012. http://www.gizmag.com/stanford-all-carbonsolar-cell/24819/.

[157] Data from the U.S. Energy Information Administration.

[158] U.S. Energy Information Administration, International Energy Outlook 2014 (DOE/EIA-0484), September 2014. http://www.eia.gov/forecasts/ieo/pdf/0484(2014).pdf.

[159] Data from U.S. Energy Information Administration.

[160] The Economist, "Seeing the Back of the Car," September 22, 2012. http://www.economist.com/node/21563280; Chris Mooney, "A Simple Guide to the Sudden Collapse in Oil Prices," Washington Post, December 1, 2014. http://www.washingtonpost.com/blogs/wonkblog/wp/2014/11/28/a-simple-guide-to-the-sudden-collapse-in-oil-prices/.

[161] Data from U.S. Energy Administration.

[162] Data from China National Bureau of Statistics, via CEIC.

[163] Data from U.S. Energy Information Administration.

[164] Data from World Bank Development Indicators.

[165] Joachim Betz, The Reform of China's Energy Policies. GIGA Working Papers 216 (Hamburg: German Institute of Global and Area Studies, February 2013); data from World Bank Development Indicators.

[166] Alex L. Wang, "The Search for Sustainable Legitimacy: Environmental Law and Bureaucracy in China," Harvard Environmental Law Review 37 (2013): 365-440.

[167] Lynn K. Price, Xuejun Wang, and Jiang Yun, China's Top-1000 Energy-Consuming Enterprises Program: Reducing Energy Consumption of the 1000 Largest Industrial Enterprises in China (Berkeley, CA: E.O. Lawrence Berkeley National Laboratory, 2008). http://eetd.lbl.gov/node/49502.

[168] International Partnership on Mitigation and MRV, "The Top 10,000 Programme." http://www.mitigationpartnership.net/gpa/implementing-national-energy-efficiency-programme-0.

[169] Nicholas Lardy, Markets Over Mao (Washington, DC: Peterson Institute for International Economics, 2014), p. 16.

[170] Paul Pennay, "China's New Coal Tax Takes Effect," Daily Telegraph, December 2, 2014.

[171] Data from CNKI China Academic Journals.

[172] Edward Wong, "As Pollution Worsens in China, Solutions Succumb to Infighting," New York Times, March 21, 2013. http://www.nytimes.com/2013/03/22/world/asia/as-chinas-environmental-woes-worsen-infighting-emerges-as-biggest-obstacle.html?pagewanted=all&_r=0.

[173] Damien Ma, "China's Search for a New Energy Strategy Time to Liberalize Energy Prices," Foreign Affairs, June 4, 2013. http://www.foreignaffairs.com/articles/139425/damien-ma/chinas-search-for-a-new-energy-strategy.

[174] Data from World Energy Council. http://www.worldenergy.org/data/resources /hydropower/.

[175] Chan Wai-Shin, "Does China Have Enough Water to Keep Building Three Power Stations a Week?" China Dialogue, October 10, 2012. https://www.chinadialogue.net/article/show /single/en/5198-Does-China-have-enough-water-to-keep-building-three-powerstations-a-week.

[176] Chan Wai-Shin, "Does China Have Enough Water to Keep Building Three Power Stations a Week?" China Dialogue, October 10, 2012. https://www.chinadialogue.net/article /show/single/en/5198-Does-China-have-enough-water-to-keep-building-three-power-stations-a-week.

[177] Chan Wai-Shin, "Does China Have Enough Water to Keep Building Three Power Stations a Week?" China Dialogue, October 10, 2012. https://www.chinadialogue.net

/article/show/single/en/5198-Does-China-have-enough-water-to-keep-building-three-power-stations-a-week.

[178] Scott Moore, "China's Massive Water Problem," The New York Times, March 28, 2013. http://www.nytimes.com/2013/03/29/opinion/global/chinas-massive-water-problem.html?_r=0.

[179] Angel Hsu (Associate Research Scientist and Lecturer, Yale School of Forestry and Environmental Studies), e-mail to Commission staff, January 21, 2015.

[180] Sophie le Clue, Geopolitical Risks: Transboundary Rivers (China Water Risk, February 2012). http://chinawaterrisk.org/resources/analysis-reviews/geopolitical-risks-transboundary-rivers/.

[181] Michael Levi, The Power Surge (Oxford: Oxford University Press, 2013), pp. 20-50.

[182] Han Wenke and Yang Yufeng, China Energy Outlook: Executive Summary (Beijing: China National Development and Reform Commission, 2012), pp. 38-40.

[183] Data from International Energy Agency.

[184] Paritosh Bansal and Farah Master, "Chesapeake, CNOOC Strike Second Shale Deal for $1.3 Bln," Reuters, January 30, 2011. http://www.reuters.com/article/2011/01/31/cnooc-chesapeake-idUKN3017000120110131.

[185] U.S. Energy Information Administration, "China." http://www.eia.gov /countries /analysisbriefs/China/china.pdf.

[186] Benjamin Haas and Rakteem Katakey, "China's Shale Gas No Revolution as Price Imperils Output: Energy," Bloomberg, February 19, 2013. http://www.bloomberg.com/news/2013-02-19/china-s-shale-gas-no-revolution-as-price-imperils-output-energy.html.

[187] Chen Aizhu, "China's Ragtag Shale Army a Long Way from Revolution," Reuters, March 10, 2013. http://www.reuters.com/article/2013/03/10/us-china-shale-idUSBRE9290 GR20130310.

[188] Melanie Hart and Daniel J. Weiss, Making Fracking Safe in the East and West (Washington, DC: Center for American Progress, October 2011): 6-7.

[189] Jane Nankano et al., Prospects for Shale Gas Development in Asia: Examining Potentials and Challenges in China and India, (Center for Strategic and International Studies, August 2012), pp. 6-7.

[190] Yue Wang, "China's Coming Shale Gas Auction Offers Little Hope to Private Investors," Forbes, August 7, 2014. http://www.forbes.com/sites/ywang/2014/08/07/chinas-coming-shale-gas-auction-offer-little-hope-to-private-investors/.

[191] Der Spiegel, "Roadmap for the Energy Revolution: Germany to Phase out Nuclear Power by 2022," May 30, 2011. http://www.spiegel.de/international/germany/roadmap-for-the-energy-revolution-germany-to-phase-out-nuclear-power-by-2022-a765594.html.

[192] Matthew L. Wald, "Approval of Reactor Designs Clears Path for New Plants, New York Times, December 22, 2011. http://www.nytimes.com/2011/12/23/business/energy-environment/nrc-clears-way-for-new-nuclear-plant-construction.html.

[193] U.S. Energy Information Administration, China Analysis Brief, April 22, 2013. http://www.eia.gov/countries/analysisbriefs/China/china.pdf.

[194] World Nuclear Association, "Nuclear Power in China." http://www.world-nuclear.org/info/country-profiles/countries-a-f/china-- nuclear-power/.

[195] Christina Larson, "China Wants Nuclear Reactors, and Lots of Them," Bloomberg Businessweek, February 21, 2013. http://www.businessweek.com/articles/2013-02-21/china-wants-nuclear-reactors-and-lots-of-them.

[196] National Nuclear Security Administration, "U.S., China Advance Nuclear Safety and Security Cooperation through Peaceful Uses of Nuclear Technology," April 24, 2012. http://nnsa.energy.gov/mediaroom/pressreleases/punt42412.

[197] World Nuclear Association, "Nuclear Power in China." http://www.world-nuclear.org/info/country-profiles/countries-a-f/china-- nuclear-power/.

[198] World Nuclear Association, "Nuclear Power in China." http://www.world-nuclear.org/info/country-profiles/countries-a-f/china-- nuclear-power/.

[199] Want China Times, "China Seeks to Export Nuclear Power Technology," November 3, 2011. http://www.wantchinatimes.com/newssubclass-cnt.aspx?id=20141103000039&cid=%201502.

[200] Evan Osnos, "Boss Rail," The New Yorker, October 22, 2012. http://www.newyorker.com /magazine/2012/10/22/boss-rail.

[201] Melanie Hart, China's Shifting Stance on Hydrofluorocarbons (Center for American Progress, June 12, 2013). http://www.americanprogress.org/issues /chinas-shifting-stance-on-hydrofluorocarbons/.

[202] Melanie Hart, China's Shifting Stance on Hydrofluorocarbons (Center for American Progress, June 12, 2013). http://www.americanprogress.org/issues /chinas-shifting-stance-on-hydrofluorocarbons/.

[203] Anna Fifield, "China and US Agree Non-Binding Climate Plan," Financial Times, July 10, 2013.

[204] The White House Office of the Press Secretary, Fact Sheet: U.S.-China Joint Announcement on Climate Change and Clean Energy Cooperation, November 11, 2014. http://www.whitehouse.gov/the-press-office/2014/11/11/fact-sheet-us-china-joint-announcementclimate-change-and-clean-energy-c.

[205] The White House, The President's Climate Action Plan, June 2013. http://www.whitehouse.gov/sites/default/files/image/president27sclimateactionplan.pdf.

[206] The White House Office of the Press Secretary, Fact Sheet: U.S.-China Joint Announcement on Climate Change and Clean Energy Cooperation, November 11, 2014. http://www.whitehouse.gov/the-press-office/2014/11/11/fact-sheet-us-china-joint-announcementclimate-change-and-clean-energy-c.

[207] Wan Xu and David Stanway, "China's Carbon Intensity Falls over 3.5 Percent in 2012: Official," Reuters, January 10, 2013. http://www.reuters.com/article/2013/01/10/us-china-carbon-intensity-idUSBRE9090I220130110.

[208] Wan Xu and David Stanway, "China's Carbon Intensity Falls over 3.5 Percent in 2012: Official," Reuters, January 10, 2013. http://www.reuters.com/article/2013/01/10/us-china-carbon-intensity-idUSBRE9090I220130110.

[209] British Petroleum, "Energy Economics." http://www.bp.com/en/global/corporate/about-bp/energy-economics.html.

[210] Michael Davidson, Gretchen Greene, and Mingming Liu, Untapped Potential: Energy Savings and Climate Benefits from Strengthening Water Use Efficiency in China's Building and Industrial Sector. Woodrow Wilson Center China Environment Series 2012/2013, p. 10.

[211] Angel Hsu (Associate Research Scientist and Lecturer, Yale School of Forestry and Environmental Studies), e-mail to Commission staff, January 21, 2015; Economist, "Browner but Greener," February 1, 2014. http://www.economist.com/news/china /21595487-chinastands-out-its-greenness-new-environmental-ranking-browner-greener.

[212] International Partnership for Energy Efficiency Cooperation Blogs, "China Is on Track to Meet its 12th Five-Year Plan Energy Intensity Target," May 19, 2014. http://www.ipeec.org/blog/view/id/797.html.

[213] Benjamin Haas, "China Emission Trading Experiment Unlikely to Ease Cities' Smog," Bloomberg, June 17, 2013. http://www.bloomberg.com/news/20 06-17/china-emission-trading-experiment-unlikely-to-ease-cities-smog.html.

[214] BBC, "China Orders Firms to Curb Pollution," June 15, 2013. http://www.bbc.co.uk/news/world-asia-china-22924231.

[215] Angel Hsu (Associate Research Scientist and Lecturer, Yale School of Forestry and Environmental Studies), e-mail to Commission staff, January 21, 2015.

[216] U.S.-China Economic and Security Review Commission, Hearing on China's Internal Dilemmas, written testimony of Elizabeth Economy, February 25, 2011.

[217] Edward Wong, "China Lets Media Report on Air Pollution Crisis," New York Times, January 15, 2013. http://www.nytimes.com/2013/01/15/world/asia/china-allows-media-to-report-alarming-air-pollution-crisis.html.

[218] Christopher K. Johnson, China Convenes the 12th National People's Congress (Center for International and Strategic Studies, March 5, 2013). http://csis.org/publication/china-convenes-12th-national-peoples-congress.

[219] BBC News, "China Blocks U.S. Air Pollution Data as Apec Leaders Meet," November 11, 2014. http://www.bbc.com/news/world-asiachina-29999784.

[220] Todd Woody, "Solar Industry Anxious over Defective Panels," New York Times, May 28, 2013. http://www.nytimes.com/2013/05/29/business/energy-environment/solar-powers-dark-side.html.

In: China's Wind and Solar Industries
Editor: Rita Schwartz
ISBN: 978-1-63483-166-6

Chapter 2

STATEMENT OF LEOCADIA I. ZAK, DIRECTOR, U.S. TRADE AND DEVELOPMENT AGENCY. HEARING ON "U.S.-CHINA CLEAN ENERGY COOPERATION: STATUS, CHALLENGES, AND OPPORTUNITIES"*

Senator Goodwin, Commissioner Cleveland and members of the Commission, thank you for the opportunity to testify about the U.S. Trade and Development Agency's cooperation on clean energy development with China. We welcome the Commission's interest in USTDA's work to foster public-private partnerships between the U.S. and Chinese energy sectors, and we look forward to describing the status, challenges and opportunities of those efforts.

USTDA's dual mandate positions our Agency to create jobs here at home, while promoting sustainable infrastructure in emerging markets around the world. USTDA was created to "promote United States private sector participation in development projects in developing and middle-income countries, with special emphasis on economic sectors with significant United States export potential."[1] USTDA is distinctive among federal agencies in that it is mandated by Congress to engage the U.S. private sector in development projects at the critical early stages when technology options and project requirements are being defined.[2] By highlighting opportunities for the use of U.S. expertise and technology when they can effectively be incorporated into

---

* This is an edited, reformatted and augmented version of a statement presented April 24, 2014 before the U.S.-China Economic and Security Review Commission.

project planning, the Agency increases opportunities for the use of U.S. exports in project implementation while helping to safeguard U.S. commercial interests.

USTDA is a streamlined, nimble agency that takes rapid and targeted action to create meaningful project-building partnerships when the need and opportunity for them are greatest. As explained by the Center for Strategic and International Studies in its 2011 report, *USTDA: Good Value for Development Dollars*, these partnerships – in addition to the Agency's other tools – give USTDA the "unique ability to leverage its assets in a multitude of ways: to strengthen the domestic economy, continue international development priorities, and serve diplomatic interests in emerging markets around the world." [3]

USTDA's Mission

USTDA's dual Congressional mandate requires that the Agency both (a) provide foreign assistance for trade and economic development and (b) help to put Americans to work in the jobs that result from exports. In the Agency's history of linking U.S. businesses to export opportunities, USTDA has generated over $45.8 billion in U.S. exports[4] and has become the leading U.S. government agency for early project development and planning activities in emerging economies. The Agency's effectiveness is demonstrated by the fact that its programs are now generating more U.S. exports per program dollar than at any other time in the Agency's history: $73 of exports for every dollar programmed, up from $41 just four years ago. In FY 2013 alone, the Agency identified $2.95 billion of new exports generated from USTDA-funded activities, which has helped support approximately 14,000 jobs in the United States.[5]

The Agency accomplishes its mission by providing grants to overseas sponsors for priority infrastructure development activities in their countries. The funding may be used to perform a feasibility study, provide technical assistance or launch a pilot project. USTDA also connects project developers with U.S. businesses through its reverse trade missions, which are specially tailored to bring foreign decision-makers to the United States to observe the design, manufacture and operation of U.S. products and services in order to inform their procurement decisions.

These activities have produced results for both U.S. industry and USTDA's partners in emerging markets: U.S. companies are provided access to the lead infrastructure project developers around the world, while foreign partners gain insight into the latest, most appropriate U.S. technologies to meet their development needs. USTDA focuses its program on sectors where U.S. firms are globally competitive, such as energy, transportation and telecommunications. As a result, the Agency is able to provide targeted foreign assistance, support U.S. trade and economic development priorities, and promote U.S. job creation.

USTDA's Collaborative Partnerships

USTDA is distinct among the U.S. government's foreign assistance agencies because, as required by its mandate, it partners with the U.S. private sector at the very early stages of project development to jointly craft solutions to developmental challenges around the world. In the course of providing these solutions, USTDA collaborates with a wide variety of U.S. government agencies and multilateral institutions in a manner that ensures success. USTDA has been recognized by its private and public sector partners for the flexibility of its program, the range of tools at its disposal and the speed with which it can deploy results-oriented assistance.

USTDA believes in the crucial importance of integrating government resources and industry innovations for win-win results. Ten years ago, this philosophy led USTDA to support the creation of a public-private partnership to promote commercial, policy and technical cooperation between the U.S. and Chinese aviation sectors. The U.S.-China Aviation Cooperation (ACP) was developed in response to requests from USTDA's partners in U.S. industry, who wanted to take advantage of opportunities in China's rapidly growing aviation market but needed the U.S. government's support in order to gain access to the key decision-makers from Chinese government and industry. The Agency also heard from its Chinese government counterparts, who were seeking technical expertise for their infrastructure development projects. USTDA successfully convened all parties to develop a framework for the ACP, which has grown to include over 40 U.S. industry members and five public sector partners since it was founded in 2004.

U.S.-China Energy Cooperation Program

After experiencing the benefits of working closely with their Chinese counterparts in a public-private aviation partnership, several U.S. industry leaders approached USTDA to propose a similar structure for energy cooperation. Government and industry representatives from both countries collaborated to develop a plan for the U.S.-China Energy Cooperation Program (ECP), which was launched during President Obama's first visit to Beijing in November 2009.

The ECP was created, in part, to connect U.S. businesses to the enormous opportunities that exist in China's $1 trillion clean technology market, as well as the significant energy investments planned by the Chinese government. For example, the Chinese government intends to invest $530 billion by 2020 on smart grid infrastructure.[6] As the world's technical leaders in dynamic subsectors like smart grid, U.S. industry can offer Chinese decision-makers goods and services that can help them achieve their ambitious goals.

The United States and Chinese governments have committed to helping facilitate this type of collaboration. During the same state visit that launched the ECP in 2009, President Obama and President Hu Jintao committed to improving energy security and combating climate change by reducing energy waste in the U.S. and China. They announced joint efforts to promote greener buildings and communities, improve industrial energy efficiency, harmonize consumer product standards and advance energy efficiency technologies.[7] Indeed, leaders in both the United States and China – which, as the world's two largest energy consumers, together account for more than 50% of global greenhouse gas emissions – recognize the importance of mitigating the effects of climate change. China is already facing significant environmental challenges, particularly in the area of air quality. As the U.S.-China Economic and Security Review Commission has noted, Beijing's reading of airborne particulates in February 2014 was 11 times the recommended exposure limit set by the World Health Organization. And in April 2013, a British medical journal published new data indicating that 1.2 million people died premature deaths in China in 2010 due to outdoor air pollution, roughly 40% of the global total.[8]

Clean energy cooperation has been a focus of USTDA's work in China since the Agency entered into a framework agreement with the Chinese Ministry of Commerce in 2001. Since that time, USTDA has partnered with China's Ministry of Environmental Protection (MEP) on over 40 environment and climate change projects. The ECP has provided a valuable forum for this collaboration by fostering creative solutions to reduce environmental impacts and increase energy efficiency. The ECP provides a platform for its 52 company members, as well as its 10 industry subsector working groups, to share information and experiences while working together to address important issues.

Additionally, the ECP often paves the way for U.S. small and medium enterprises (SMEs) seeking entry into China's market by providing them with greater visibility and additional resources. While big companies have large overseas offices complete with government relations staff, smaller companies rarely have the resources to represent their interests as strongly. Cooperation programs like the ECP provide SMEs access to important decision-makers and resources; the American Chamber of Commerce in Shanghai, for example, which has a dedicated SME Center, has supported several ECP and USTDA activities in the region.

In addition, public-private partnerships frequently give SMEs a voice in policy decisions. For instance, USTDA has supported several of the ECP's small- and medium-sized firms in encouraging China to adopt U.S.-style standards so their products could compete in the Chinese market. One of the ECP's founding members was Solatube, a small California-based company with patented daylighting devices that use advanced optics to bring natural lighting into interior building spaces. When Solatube first arrived in China, the daylighting industry was non-existent. Because of the absence of a market and associated standards, it was clear to the company's leaders that establishing reasonable standards would help them generate acceptance of – and create demand for – their high-quality systems.

Through the ECP, Solatube participated in two USTDA reverse trade missions during which Chinese delegations visited Oregon and California, where they saw demonstrations of the company's daylighting technologies. Following the first visit, Solatube secured its first government project in Henan Province for the design and installation of daylighting units in two buildings. Solatube then won a 20,000-square meter project in Shanxi Province. They were also invited to participate in the drafting and revision of three building lighting standards that now include daylighting systems. In response, Solatube's General Manager remarked that the "ECP gives small and medium enterprises, especially those leading a new industry, the leverage needed to work with the government, and to reach out to the Chinese market and get results."

The following examples illustrate the ECP's value in helping to mitigate China's short- and long-term environmental issues, and in connecting U.S.

companies to the opportunities that exist in China's rapidly expanding energy market.

Air Quality and Environmental Improvement Efforts

The ECP has undertaken a number of projects to improve China's environment since its inception – building on the last decade of work by USTDA, which has invested nearly 20% of its East Asia portfolio on activities supporting clean water and the environment in China. Both before and after the launch of the ECP, USTDA has sponsored activities designed to develop environmental early warning and emergency response command systems; improve energy efficiency across the construction supply chain; and reduce emissions of sulfur dioxide, nitrogen oxide and particulate matter, as well as mercury and other heavy metals.

For example, USTDA funded a reverse trade mission in 2012 designed to introduce Chinese regulators and power providers to U.S. mercury emissions control technologies and regulatory best practices. This activity has allowed USTDA and the ECP, along with the U.S. Environmental Protection Agency (EPA) and Department of Commerce, to enter into a dialogue with the Chinese government about the social, environmental and economic benefits of setting stringent mercury emission standards for Chinese power plants. Led by ECP members, this dialogue is highlighting U.S. technologies and best practices that can be applied in China's power-plant sector.

The focus on mitigating the environmental impacts of the country's energy consumption has become increasingly important to the Chinese government, which has announced significant emissions reduction goals. China's Twelfth Five-Year Guideline, which was released in 2010, introduced for the first time an emissions target to reduce carbon intensity (CO2 per unit of GDP) to 17% below 2010 levels by 2015.[9] More recently, the Chinese government has added a new goal of reducing average levels of PM2.5 particulates in 47 cities by 5% based on 2010 levels before 2015.[10]

USTDA and the ECP are currently supporting efforts to help China meet these ambitious goals. The China Air Quality Management Program, developed in cooperation with the EPA, will assist China's MEP in meeting its efforts to reduce pollution across the country by developing an air quality management plan based on U.S. best practices. The project will include a series of workshops to educate the plan's developers on U.S. industry-leading clean air technologies. And this year, USTDA and the ECP will host a reverse trade mission to assist China's efforts to develop comprehensive regulatory frameworks, methodologies and solutions to provide clean, healthy air for its population.

Smart Solutions for Electricity Transmission and Distribution

The Smart Grid Working Group, which was the first Working Group to emerge under the ECP, quickly became the largest and remains so today. It was founded by information technologies (IT) companies that could offer solutions to improve electricity transmission and distribution within China's power grid – an important priority for the government of China, which surpassed the U.S. as the world's largest smart grid market in 2013.[11] Both the U.S. companies and the Chinese leadership understood that, compared to traditional electricity generation investments, smart grid investments can be realized much quicker and at much lower costs, resulting in savings that put less demand on utilities for the supply of new power generation. Equally important, smart grid investments also provide a larger pathway for utilities to receive power from renewable energy sources.

Driven by industry leadership, USTDA and the ECP have collaborated on nearly a dozen electricity transmission and distribution activities in China, including feasibility studies, pilot projects and technical workshops and trainings. For example, USTDA supported a pilot project with ECP members Honeywell (Morristown, NJ) and AECOM (Los Angeles, CA) to help the China State Grid Corporation's Electric Power Research Institute achieve its demand side management reduction goals. The recently-completed pilot demonstrated that U.S. technology could reduce energy loads of commercial buildings in China by 15% and that of industrial sites by as much as 50% – highly significant savings. China State Grid Corporation is planning to expand this pilot to other cities, presenting follow-on contract opportunities for Honeywell, which won the Chinese *Economic Observer*'s 2013 China Low Carbon Model Award for their contributions to the project.

Building on this effort, USTDA partnered with China's National Energy Administration (NEA) and the U.S. Federal Energy Regulatory Commission (FERC), as well as several ECP members, to host a successful smart grid technologies conference in Shenzhen in 2012. The conference attracted almost 300 U.S. and Chinese attendees interested in sharing best practices about grid interconnectivity and optimization.

To solidify the discussions undertaken by the participants at the event, USTDA supported follow-on projects that will advance the ECP's and China's efforts to deploy smart grid technologies, including pilot projects for fuel cell and communications applications. For example, USTDA funded a smart grid substation communication architecture pilot project. Led by Cisco (San Jose, CA) in cooperation with the China Electric Power Research Institute, the project is demonstrating U.S. technologies for smart grid communication at the substation level, which will greatly enhance the overall reliability of China's electrical grid. This project will pave the way for introduction of state-of-the-art integrated communications capability at the 8,100 Chinese substations being upgraded to smart substations over the next two years.

Standards Development for Vehicle Emissions

The ECP's Clean Transportation Working Group recently requested USTDA's assistance in its efforts to address another major source of China's emissions: mid-range and heavy duty commercial vehicles. Over the last decade, China's reliance on rail for goods transportation decreased by 30% in favor of road freight. And in 2012 alone, China sold almost 4 million commercial heavy duty vehicles (HDVs) – nearly seven times the size of U.S. sales. This upward trend in commercial HDVs is having a significant effect on both fuel consumption and air emissions in China; according to the Heavy Duty Manufacturers Association, "commercial vehicles account for only 20% of the total motor vehicles running in China but consume nearly half of the fuel needed by all vehicles in use in the country."[12]

In response to the Working Group's request, USTDA led an effort with the U.S. Department of State to align U.S. and Chinese regulations on fuel economy standards for HDVs. This will help China develop policies that reduce greenhouse gases and improve fuel consumption, and will also help level the playing field for trade.

As part of this effort, USTDA brought delegates from China's MEP, Ministry of Industry and Information Technology, Automotive Technology and Research Center, and Internal Combustion Engine Industry Association to the United States in December 2013. The reverse trade mission was designed to introduce Chinese delegates to U.S. best practices in regulations for HDV fuel economy standards, as well as to showcase advanced U.S. technologies that can assist China in meeting fuel economy and emissions requirements. U.S. firms manufacture most of the world's high-end technology for HDVs, such as engines, advanced machining, powertrain controls and fuel emissions control equipment.

While it is too early to determine U.S. exports as a result of this activity, it has already made important strides toward harmonizing fuel economy standards in China. And it demonstrates the values of public-private partnerships like the ECP, which achieve results by convening the right players to address a common issue or challenge. The outcome is dynamic engagement, mutual benefit and long-term relationship-building.

USG Interagency Collaboration

Several U.S. government technical agencies have contributed to the success of the ECP since its launch. In fact, the Departments of Energy and Commerce joined USTDA in signing the Memorandum of Understanding that officially formed the ECP in 2009. Along with other USG stakeholders like the Department of State, FERC and the EPA, these agencies continue to support the ECP's efforts to connect Chinese decision-makers to U.S. technical expertise in clean energy. This exemplary interagency collaboration has been highlighted at the U.S.-China Strategic and Economic Dialogue, the U.S.-China Joint Commission on Commerce and Trade, and other events.

CONCLUSION

USTDA's unique and targeted focus on engaging the U.S. private sector in early project planning for development projects places it at the forefront of an innovative, sustainable foreign assistance model. The Agency moves quickly to build mutually beneficial partnerships between the U.S. private sector and overseas project sponsors at a critical point in project planning. The Agency's public-private partnerships, particularly the U.S.-China Energy Cooperation Program, are viewed as touchstone programs that allow government and industry partners to accomplish more together than they could separately. In fact, when he addressed the ECP for the first time last April, Secretary Kerry called the Program "the best of international and government-to-government cooperation."[13] The primary value of the ECP – and of USTDA's program in general, which has generated almost $300 million in U.S. exports for energy and environmental projects in China since 2001 – is its ability to identify tangible business opportunities for U.S. companies while contributing to China's clean energy development and emissions reduction efforts.

As an added benefit, the ECP's stakeholders have also observed that their work can positively impact the priorities, standards and policies of the Chinese government in the energy sector. Through their collaboration with the ECP, Chinese decision-makers have begun to look to U.S. companies for affordable technological solutions to their development challenges. And, as demonstrated by the examples of Solatube and the dialogue on the reduction of mercury emissions, the Chinese have begun to introduce the policies and standards that can enable the implementation of U.S. clean energy technologies.

End Notes

[1] 22 U.S.C. § 2421(a).

[2] 22 U.S.C. § 2421(b)(2).

[3] Daniel F. Runde and Lauren Bieniek, USTDA: Good Value for Development Dollars, Center for Strategic and International Studies, October 21, 2011, available at https://csis.org/publication/ustda-good-value-development-dollars, last accessed April 2014.

[4] This historic cumulative export total includes data collected by the Agency's predecessor organization, prior to USTDA's formation as an independent agency in 1992.

[5] Martin Johnson and Chris Rasmussen, Jobs Supported by Exports 2012: An Update, Office of Competition and Economic Analysis, International Trade Administration, Department of Commerce, February 26, 2013, available at http://www.trade.gov/mas/ian/build/groups/public/@tg_ian/documents/webcontent/tg_ian_004021.pdf, last accessed April 2014.

[6] China Greentech Initiative, The China Greentech Report 2011: China's Emergence as a Global Greentech Market Leader, pg. 16, available at http://cgtr.china-greentech.com/ China GreentechReport2011-final.pdf, last accessed April 2014.

[7] The White House, "FACT SHEET: U.S.-China Energy Efficiency Action Plan," November 17, 2009, available at http://www.whitehouse.gov/files/documents/2009/november/US-China-Fact-Sheet-on-Efficiency-Action-Plan.pdf, last accessed April 2014.

[8] Iacob Koch-Weser, USCC Economic Issue Brief No. 3, "China's Hunger for U.S. Planes and Cars: Assessing the Risks," March 27, 2014, available at http://origin.www.uscc.gov/sites/default/files/Research/USCC%20Economic%20Issue%20Brief_ChinasHunger_03.27.14.pdf, last accessed April 2014.

[9] China Greentech Initiative, China Greentech Report 2013: China at a Crossroads, pg. 32, available at http://report.china-greentech.com/en.html, last accessed April 2014.

[10] Ibid, pg. 21.

[11] Bloomberg New Energy Finance, "China Outspends the US for the First Time in $15BN Smart Grid Market," February 20, 2014, available at http://about.bnef.com/press-releases/china-out-spends-the-us-for-first-time-in-15bn-smart-grid-market/, last accessed April 2014.

[12] Heavy Duty Manufacturers Association, "China MIIT to Set Fuel-Consumption Limits for Heavy-Duty Commercial Vehicles," September 27, 2012, available at http://www. hdma. org/Main-Menu/HDMA-Publications/Diesel-Download-2013-Archive/Intl-Diesel-

Download/September-27-2012/China-MIIT-to-Set-Fuel-Consumption-Limits-for-Heavy-Duty-Commercial-Vehicles.html, last accessed April 2014.

13 Secretary of State John Kerry, "Remarks at Energy Cooperation Event," Beijing, China, April 13, 2013, available at http://www.state.gov/secretary/remarks/2013/04/207474.htm, last accessed April 2014.

In: China's Wind and Solar Industries
Editor: Rita Schwartz
ISBN: 978-1-63483-166-6

Chapter 3

TESTIMONY OF JOANNA LEWIS, ASSISTANT PROFESSOR OF SCIENCE, TECHNOLOGY AND INTERNATIONAL AFFAIRS, GEORGETOWN UNIVERSITY. HEARING ON "U.S.-CHINA CLEAN ENERGY COOPERATION: STATUS, CHALLENGES, AND OPPORTUNITIES"*,[1]

Members of the Commission: good morning, and thank you for the invitation to participate in this hearing to discuss U.S.-China Clean Energy Cooperation. I am currently a professor at Georgetown University, and have been working on the issue of U.S.-China clean energy cooperation for several years in a variety of roles. My own academic research focuses on the evolving nature of U.S.-China relations on energy and climate, including models of and obstacles to clean energy cooperation and to clean energy technology transfer, as well as China's domestic energy and climate policy strategy. Currently I am the Principal Investigator on a multi-year National Science Foundation supported research project, "International Partnerships and Technological Leapfrogging in China's Clean Energy Sector," in which I'm conducting an international review of bilateral clean energy cooperation initiatives with China.

Previously I served as the Research Director of the Asia Society's Initiative for U.S.-China Cooperation on Energy and Climate, and led the

---

* This is an edited, reformatted and augmented version of testimony presented April 24, 2014 before the U.S.-China Economic and Security Review Commission.

drafting of a "Roadmap for U.S.-China Cooperation on Energy and Climate Change" in 2008 that aimed to inform the incoming Administration's clean energy cooperation priorities with China. From 2008-2010 I served on a panel of the U.S. National Academies to review "U.S.-China Cooperation on Electricity from Renewable Sources" along with the Chinese Academies of Sciences and Engineering. I have also been directly involved in many U.S.-China clean energy cooperation projects myself, including as a researcher at Lawrence Berkeley National Laboratory's China Energy Group, as an international advisor to the Energy Foundation's China Sustainable Energy Program, and as a visiting scholar at Tsinghua University.

As the largest energy consuming nation and largest greenhouse gas emitter in the world, China plays an increasingly important role in the geopolitics of energy as well as climate change mitigation. A longtime proponent of multilateralism, China has been increasing its engagement and its seniority in various multilateral forums, including the United Nations.

It has, however, been viewed by many industrialized nations as an obstructionist player in the ongoing international negotiations under the United Nations Framework Convention on Climate Change. China, for many reasons, plays a very different role in the multilateral context from the role it plays in a bilateral one.

In these negotiations, China time and again has served the role of spokesperson for the developing world, defending its right to emit in the name of economic development. In a direct bilateral discussion with a leading industrialized nation like the United States, however, China wants to be seen as an equal, and as the global superpower that it has become. For this reason direct bilateral discussions can lead to more effective platforms for cooperation on topics that become politicized in a larger negotiation.

Bilateral cooperation allows for the identification of clear technical areas of mutual interest between the countries involved. For example, while China and the United States frequently disagree over core issues in the UN climate negotiations, the technical challenges behind their mitigation options are distinctly similar. Both countries have abundant domestic coal resources that provide energy security benefits but are a significant source of emissions.

While both China and the United States also have excellent renewable resources, including wind and solar, the best resources and locations for renewable power development tend to be located far from population centers and electricity demand and therefore will require expanded and modernized transmission infrastructures. Both countries have realized the potential energy efficiency gains they can achieve but still lag behind Europe, Japan, and others in developing a more efficient energy system.

China and the United States have a somewhat unique positioning in the international community on energy and climate change issues, as the two largest economies, the two largest energy consumers, and the two largest greenhouse gas emitters on the planet.

Owing to the similarities in energy systems shared by the two countries, there are many areas where both the United States and China could benefit from cooperation on climate change and clean-energy development.

The United States and China have a long history of bilateral clean energy cooperation, both through official governmental channels and between universities and nongovernmental organizations. For this testimony I am focusing specifically on official bilateral cooperation activities facilitated by the governments of the Untied States and China.

The U.S.–China Agreement on Cooperation in Science and Technology (S&T Agreement) was signed in 1979, soon after the normalization of diplomatic relations between the two countries. This agreement established a framework for many of the subsequent agreements on energy-related cooperation between the United States and China signed over the next thirty years. Many of the agreements signed during the 1970s and 1980s focused on promoting collaboration and understanding in basic research related to many core energy technologies of this period, including the 1970 Agreement on High Energy Physics, the 1983 Protocol on Nuclear Physics and Magnetic Fusion, and the 1985 Protocol on Cooperation in the Field of Fossil Energy Research and Development.

Cooperation on basic research topics continued during the 1990s, but a new set of cooperation agreements were signed focusing more on energy policy discussions. These included policy discussions on fossil and nuclear energy, energy efficiency, and renewable energy. In 1995 U.S. Secretary of Energy Hazel O'Leary signed several bilateral agreements with China on high-level energy policy consultations, nuclear energy, renewable energy, energy efficiency, coal-bed methane, and climate research. That year the initial Protocol for Cooperation in the Fields of Energy Efficiency and Renewable Energy Technology Development and Utilization was signed; it was subsequently amended with new programs of cooperation (through annexes to the agreement) in the years following.

The agreements signed from 2000 to 2010 built upon the agreements of the previous two decades, which allowed for a further broadening of the energy topics being discussed bilaterally, as well as further linkages to other bilateral discussions taking place between the United States and China on economic and security issues. In 2006 the U.S.–China Strategic Economic Dialogue (SED) was founded by Vice Premier Wu Yi and U.S. Treasury Secretary Henry Paulson. The biannual, cabinet-level dialogue involved several agencies, including the U.S. Department of Energy and Environmental Protection Agency, and China National Development Reform Commission and Ministry of Science and Technology, and included a specific track for energy and environment discussion. In 2008 the fourth SED led to the establishment of the U.S.–China Ten-Year Framework for Cooperation on Energy and Environment. The Framework involves multiple U.S. and Chinese government agencies that work on environment and energy. It initially established five joint task forces on the five functional areas of the framework: (1) clean, efficient, and secure electricity production and transmission; (2) clean water; (3) clean air; (4) clean and efficient transportation; and (5) conservation of forest and wetland ecosystems. These five areas were further elaborated in seven specific action plans for implementation and were later expanded on in the July 2009 Memorandum of Understanding to Enhance Cooperation on Climate Change, Energy, and Environment, and meetings and negotiations under this agreement continue today.

Over the past 5 years, clean energy has emerged as a leading topic of cooperation between China and the United States. In November 2009, the United States and China signed 7 new agreements on clean energy cooperation, covering energy efficiency, renewable energy, electric vehicles, advanced coal technologies, and shale gas. These agreements also launched the U.S. China Energy Cooperation Program (ECP), which helps U.S. companies engage in China's clean energy sector, and the U.S.-China Clean Energy Research Center (CERC), which facilitates joint research and development on clean energy technology by teams of scientists and engineers from both countries. While these agreements were by no means the first agreements with China to focus on clean energy, the launch of a comprehensive package of clean energy agreements at the Presidential level signified a new model, as well as an elevated status bilaterally, for U.S.-China Clean Energy Cooperation.

Five years later, this cooperation has not waned, but rather is being strengthened. In 2013, the Obama Administration worked with the incoming Xi Administration to sign several new agreements building on the package of bilateral cooperation signed back in 2009. Key among these agreements were the establishment of a high level Climate Change Working Group, which included five new initiatives: (1) Emission Reductions from Heavy Duty and

Other Vehicles, (2) Smart Grids, (3) Carbon Capture Utilization and Storage, (4) Collecting and Managing Greenhouse Gas Emissions Data, and (5) Energy Efficiency in Buildings and Industry.

The signing of such agreements is often an accomplishment in itself. For example, many issues on which the United States now has official bilateral agreements with China longeluded U.S. policymakers, including agreements that addressed somewhat sensitive topics like measuring greenhouse gas emissions data, or that targeted challenging technical and political topics like carbon capture and sequestration. However, now that many of these agreements have been underway for several years, we can begin to assess what is working well, and what could be improved.

Much is working well. For example, the U.S.-China Renewable Energy Partnership has hosted multiple industry forums in China and the United States to facilitate new collaborations. The U.S. Department of Energy reports that a recent achievement of the partnership is the signing of a MOU between the U.S. company BrightSource Energy and several Chinese entities to deploy BrightSource's concentrating solar thermal power (CSP) technology in Qinghai province. This project, valued at $350 million, is structured such that key components would be manufactured in the United States, including some which may contain sensitive intellectual property, and the technology would be demonstrated in China, serving as China's first commercial-scale deployment of CSP technology.

Another agreement, The U.S.-China Energy Efficiency Action Plan has helped facilitate multiple research and commercial collaborations between U.S. and Chinese partners, as well as policy development in China. For example, with assistance from the U.S. Department of Energy, China enacted its first-ever energy code for rural residential buildings in May 2013. DOE estimates this could save up to 50 percent of the energy used in residences that house 700 million people, in a footprint equal to the entire U.S. residential building sector.

One bilateral initiative in particular that has moved beyond negotiated text and into the realm of real cooperation is the U.S.-China Clean Energy Research Center (CERC). The goals of the CERC are to spur innovation of clean energy technologies, diversify energy supply sources, improve energy efficiency, accelerate the transition to a low-carbon economy, and help to avoid the worst consequences of climate change. The focus on innovation through joint R&D and particularly the emphasis on the creation of intellectual property makes the CERC is unique from previous U.S.-China clean energy cooperation agreements. Approximately 1100 researchers in the U.S. and

China are supported by the work of the CERC, and each of the partner organizations are either contributing funds or directly performing research.

The CERC is governed by a steering committee that includes ministerial- or secretary-level oversight from the relevant government agencies. From the United States, the lead agency is the Department of Energy (DOE). In contrast, China has 3 different government ministries all playing a leadership role in the CERC, including the Ministry of Science and Technology (MOST), the National Energy Administration (NEA), and the Ministry of Housing and Urban and Rural Development (MOHURD).

The CERC has three technology areas that were targeted for initial cooperation activities: advanced coal technologies, efficient building technologies and clean vehicle technologies.

The CERC currently consists of 88 individual projects within these three tracks, and almost all of these projects are joint collaborations between U.S. and Chinese researchers.

The types of projects that are included in the CERC are varied along the technology research, development, demonstration and deployment (RDD&D) continuum, ranging from basic science research to technology demonstration. In addition, several projects focus on policy analysis to support the technologies being developed.

In order to address intellectual property concerns related to the CERC's R&D activities head-on, each CERC consortia has agreed upon and signed a contract that details the IP rules for participation called a Technology Management Plan (TMP). The TMP was established after months of negotiations between U.S. and Chinese lawyers and the respective government agencies involved. All participants involved in the CERC's activities are subject to the provisions of the TMP, and any new participants that join the CERC consortia must agree to its terms.

The TMP was specifically designed to clarify the joint ownership of IP resulting from joint research activities, and invented jointly by signatories to the CERC protocol from both the U.S. and China. If project IP is invented by signatories from one territory only, then the TMP requires that participants agree to negotiate in good faith terms of a nonexclusive license to the participants from the other territory. There are also provisions in the TMP that encourage the sharing of data and information related to the project work with the public, except when there is a need to preserve confidentiality. If any disputes over IP arise in the context of CERC activities, there are provisions in the TMP for how they are to be resolved.

CERC participants with IP-related disputes are first supposed to try to work out a mutually agreeable resolution. If such a resolution cannot be reached, then an arbitral tribunal in accordance with the applicable rules of international law as set by the United Nations Commission on International Trade Law (UNCITRAL) is to be utilized. The inclusion of such a provision for dispute resolution is rare among collaborative research efforts, which typically leave any disputes to the individual laws of the relevant countries to resolve.

The CERC begins to shed some light on the question of how IP can be better managed to promote cross-national technology cooperation in the clean energy sector. While the CERC is still in its early stages, my own analysis of the CERC's IP framework highlights some of the unique characteristics of the model for collaborative clean energy research that it has established, as well as a range of expectations about what the CERC can and will achieve.

Many CERC participants reported that they had initially joined the initiative because they believed the IP framework could be beneficial to their continued work in China, although they did not expect that the TMP would solve all of their IP challenges. Several private companies involved in the CERC mentioned tangible results from their participation, including new business ventures, and new IP. By mid-2013, all of the CERC consortia had reported inventions and IP that originated from CERC R&D initiatives.

Almost all of the U.S. commercial participants across the three consortia mentioned that one of the biggest advantages of participating in the CERC was to gain leverage for technology demonstration projects. Many have invested their own money in the collaborations taking place under the CERC far in excess of government support because government involvement provided leverage for project approvals, and many CERC collaborations were perceived to have current or future commercial value.

While it is too early to comprehensively assess the efforts of the CERC, it is increasingly evident that the CERC provides a model for collaborative clean energy research, development and demonstration (RD&D) that is unique in the history of U.S.-China collaborations in this area. Spanning the public and private sectors and involving top researchers from universities and national laboratories in both countries, the CERC has also been credited with propelling numerous other clean energy collaborations, including some with commercial value. The scale of the CERC is extremely impressive, and it is clearly building important clean energy-focused research partnerships between China and the United States.

The CERC experience provides some useful insight about improvements that are still needed in order to effectively navigate IP concerns specific to U.S.–China collaborations.

There is still clearly a need to better alleviate the concerns of the participants related to IP protections, as well as to better educate researchers about how to use legal tools like the TMP to better facilitate collaborative research endeavors. Rather than attempt to navigate complicated IP issues, researchers not trained in IP law are more likely to avoid IP negotiations, even at the expense of potentially valuable cross- border collaborations, unless they are confident that they have institutional support. As the first U.S.–China program targeting clean energy R&D, the CERC may ultimately play an important role in building trust among the consortia participants, which could lead to even more constructive collaborations in the future.

U.S.-China clean energy bilateral initiatives provide important channels for technical cooperation and information. Without sustained support, and continued attention to IP concerns, it will be even harder for China and the United States to make progress towards developing true cross-national collaborations which ultimately could produce considerable global benefits, particularly in the clean energy field. Just because we have a record number of bilateral agreements on the books, does not mean we can reduce our focus on the relationship. The significance of the U.S.-China clean energy relationship merits a sizable and sustained level of effort.

End Note

[1] This testimony draws from Lewis, Joanna I. (2013), "Engaging China on Clean Energy Cooperation," Chapter 7 in Green Innovation in China: China's Wind Power Industry and the Global Transition to a Low-Carbon Economy. New York: Columbia University Press; and Lewis, Joanna I. (2014), "Managing intellectual property rights in cross-border clean energy collaboration: The case of the U.S.–China Clean Energy Research Center." Energy Policy 69: 546–554.

In: China's Wind and Solar Industries
Editor: Rita Schwartz
ISBN: 978-1-63483-166-6

Chapter 4

TESTIMONY OF SARAH FORBES, SENIOR ASSOCIATE, WORLD RESOURCES INSTITUTE. HEARING ON "U.S.-CHINA CLEAN ENERGY COOPERATION: STATUS, CHALLENGES, AND OPPORTUNITIES"*

Thank you for the opportunity to contribute to the deliberations of this commission. My name is Sarah Forbes, and I am a Senior Associate in the Climate and Energy Program at the World Resources Institute (WRI). WRI is a non-profit, non-partisan think tank that focuses on the intersection of the environment and socio-economic development. We go beyond research to put ideas into action, working globally with governments, business, and civil society to build transformative solutions that protect the earth and improve people's lives. We operate globally because today's problems know no boundaries. We provide innovative paths to a sustainable planet through work that is accurate, fair and independent.

I am delighted to speak with you today about clean energy cooperation between the United States and China. I would like to open my testimony with a quote from the April 2013 Joint US-China statement on Climate Change:

> "[F]orceful, nationally appropriate action by the United States and China – including large scale cooperative action – is more critical than ever. Such action is crucial both to contain climate

* This is an edited, reformatted and augmented version of testimony presented April 24, 2014 before the U.S.-China Economic and Security Review Commission.

> change and to set the kind of powerful example that can inspire the world"

In my testimony I will describe the context for US-China collaboration on clean energy, outline the need for policies that encourage innovation throughout the value chain, describe how collaboration with China can advance US energy goals, and, perhaps most importantly, give recommendations for how US-China collaboration on clean energy could be improved so that we can inspire the world and speed the transition towards a clean energy future. My testimony will discuss not only clean energy collaboration, but will also touch on the issues you raised in inviting me around the state of cooperation on shale gas.

I will expand on the following five recommendations throughout my testimony. Future US-China collaboration on clean energy and shale gas should include:

1. Creation of collaborative networks that include supporting innovators' needs and building capacity in the workforce;
2. Continued high-level strategic engagement between the US federal government and China's central government on clean energy collaboration;
3. Substantive work on developing and implementing environmental regulations for air, water, and climate impacts;
4. Private sector engagement and investment, personnel training, and more transparency; and
5. Facilitation of opportunities for joint R&D.

Context for Clean Energy Cooperation between the US and China

Many parallels exist in the US and Chinese energy profiles. Both are continent-sized countries with geographically dispersed energy resources and with energy demand centers (cities) that are often far from energy supplies. Both countries currently rely heavily on fossil fuels to power their economies, primarily drawing on coal, natural gas, and imported oil (the US and China have similar levels of crude import dependence). Both countries seek to increase energy independence by diversifying the energy mix and ramping up domestic energy production, particularly unconventional fossil fuels such as shale gas as well as renewable energy technologies.

Collaboration on clean energy – energy efficiency, renewable energy and carbon capture and storage (CCS) – and shale gas falls into these broad categories:

- Business-to-Business
- Government-to-Government
- Researcher-to-Researcher

The following provides a brief overview of the scope of existing collaboration on clean energy and shale gas, with the shale gas discussion adapted from a paper I recently drafted for the Brookings Institution[1].

BUSINESS-TO-BUSINESS (B2B)

The following are examples of key B2B collaborations[2]:

- Duke Energy and Huaneng have been cooperating on clean energy technology since 2009. The two companies have been sharing information, especially on carbon capture, utilization and storage (CCUS). In 2012, the original agreement was expanded to examine the feasibility of brining carbon capture and storage technology that had been pilot tested in China to US power plants.
- GE and State Grid Corp have formally agreed to cooperate to jointly develop standards for smart grids in China. GE has formed an equally-owned joint venture in China to offer leasing services to the energy industry and an equally-owned joint venture to market and manufacture grid monitoring and diagnostic products.
- A consortium of US and Chinese companies including Boeing, Honeywell, PetroChina, and Air China has been collaborating to develop biofuels for passenger jets. In 2011, this effort led to a successful Chinese test flight of a Boeing 747 using a 50% blend of traditional jet fuel and the new biofuel. This collaboration benefits from being a part of the US-China Energy Cooperation Program.

In terms of shale gas, Chinese companies currently possess the ability to drill wells horizontally and have some experience with hydraulic fracturing but are in the nascent stages of acquiring experience and applying these techniques to shale gas extraction. Operators and service providers in the United States currently have substantial experience with drilling and fracking shales, having mastered these techniques effectively to maximize output. Given this level of technical and experiential asymmetry, China has made it possible for US entities to strike up partnerships with Chinese companies, primarily the National Oil Companies (NOCs). But China has prohibited foreign companies from fully entering the onshore energy production sector on their own, forcing them to form partnerships with Chinese entities. As a result, several foreign companies have already begun participating in shale gas development in China, through joint ventures (JVs) such as PetroChina-Shell and China National Offshore Oil Company-British Petroleum (CNOOC-BP).[3] Leading US service providers like Schlumberger, Baker Hughes, and Halliburton also have well-established offices and/or research institutes in Beijing and provide services to Chinese companies, including the NOCs and other smaller new entrants.

China has also invested in shale gas development in the United States, with NOCs establishing JVs with US companies. These generally take the form of acquiring stakes in company assets—investments in specific shale gas plays—and not as investments in the US companies themselves. It is a model that the global oil and gas industry has long used to sustain growth and hedge against financial risk. The rationale for these Chinese investments varies. In some cases, Chinese petroleum engineers are able to spend time onsite learning about shale gas development first hand. However, limits on "access to technology" remain an oft-cited constraint on China's future shale gas developments. What is meant by "access to technology" may be best described as "know how." That is, the technology for horizontal drilling or fracking may not be the key barrier for China *per se*—instead, it is the lack of experience in applying these tools to different geological formations to maximize the flow of the gas.

Government-to-Government (G2G)

In 2009, the following key forums for G2G collaboration on clean energy were established:

- The Clean Energy Research Center, or CERC (electric vehicles, building efficiency, and advanced coal technology), is a unique bilateral platform to allow both countries' researchers to work closely on the same tasks and is operated as a public-private partnership on both sides.
- US-China Energy Cooperation Program is a government-led collaboration that includes private-sector partners on both sides, supporting projects in the areas of smart grid, green buildings, combined heat and power and renewable energy.
- US-China Renewable Energy Partnership includes road mapping renewable energy deployments in each country, an annual US-China renewable energy forum and sharing best practices.
- US-China Electric Vehicles Initiative supports joint standards development and demonstrations in a dozen cities.
- US-China Energy Efficiency Action Plan supports officials in both countries to develop codes and rating systems and benchmark industrial energy efficiency as well as a training program for inspectors and an annual US-China Energy Efficiency forum.

In 2013, the Climate Change Working Group (CCWG) for the Strategic and Economic dialogue completed a report that took stock of the existing collaboration. The CCWG was established in April 2013 to deepen and extend existing collaboration that spurs large-scale cooperative efforts. The report resulted in a July 2013 commitment to new action initiatives for collaboration on five issues, which will draw on the relevant agencies in each country:

1. Emissions reductions from heavy-duty and other vehicles
2. Smart grid
3. Carbon capture utilization and storage
4. Collecting and managing GHG emissions data
5. Energy efficiency in buildings and industry

With respect to shale gas specifically, in 2009, President Obama and then-President Hu Jintao announced the launch of the US-China Shale Gas Resource Initiative, with the goal of sharing information about shale gas exploration and technology to reduce greenhouse gas emissions, promote energy security, and create commercial opportunities. Activities conducted under the initiative include forums, workshops, and a Chinese delegation's visit to a US shale gas development operation. The governments' collaboration includes the following activities, led by key agencies:

- The US Geological Survey and the Department of Energy (DoE) have worked with Chinese counterparts to develop estimates for China's shale gas resource. Although the collaboration resulted in sharing information on methodology, Chinese geological data is considered a state secret and cannot be shared with foreigners.
- DoE manages the US-China oil and gas industry forum, which sponsors an annual meeting designed to bring industry players together to share information via technical presentations.[4] In September 2012, the forum sponsored a meeting focused on shale gas. DoE also has relevant work underway that focuses on issues under Annex III of the bilateral Fossil Energy Protocol.[5]
- In April 2013, the US Trade and Development Administration partnered with the National Energy Administration on a training program. The program included four short courses led by the Gas Technologies Institute and targeted attendees from the Chinese government and industry.[6]

Researcher-to-Researcher (R2R)

Regardless of the G2G or B2B framework for collaboration, the effectiveness of collaboration on clean energy technology development ultimately depends on the relationships established between researchers. G2G frameworks like the CERC are designed to foster such R2R relationships, however the depth and extent of collaboration ultimately depends on a collection of individual relationships among researchers who share the same interests and vision.

It is increasing common for effective G2G, B2B and R2R collaborations to result in bilateral public/private partnerships:

- CERC participants from Yanchang Petroleum, University of Wyoming, and the China Academy of Sciences are partnering on a CCUS/Enhanced Oil Recovery project in Shaanxi province.
- CERC participants from Huaneng Energy, Lawrence Livermore National Laboratory, and Duke Energy have conducted carbon

capture cost modeling on Duke's Gibson Power Plant and Huaneng's Shidonku plant.

FUTURE ENVIRONMENTAL CHALLENGES REQUIRE INNOVATION

Transforming the global energy picture requires innovation. Such innovation will not happen without investment from the public and private sector, commitment from governments and industry, support from local communities, and out-of-the box ideas that push the envelope of what seems possible. All of these must be sustained over time.

R&D spending alone is not effective in deploying new energy technologies, and the innovations created in U.S. national laboratories and universities often do not get manufactured or deployed into the marketplace. For instance, a recent WRI report comparing clean energy industries across major countries highlighted that the United States had the highest public investments in wind energy R&D. Yet, it was the only wind energy market among the five countries analyzed that maintained a long-term trade deficit in wind equipment, importing more than it exports, due largely to the uncertainty surrounding the longevity of policies that support deployment of wind energy technologies.[7]

WRI's work on innovation concluded the following:[8]

> "Innovation is a powerful, cumulative process but it does not happen automatically in a highly regulated sector like electricity. It is critical that policymakers support innovators by building a robust, dynamic innovation ecosystem. This goes beyond investing in public research and development and creating markets through subsidies. It also includes building collaborative networks, creating stable regulatory environments, providing infrastructure, supporting innovators' needs for finance, and building capacity in the workforce."

International collaboration, and US-China collaboration specifically, is one way to create collaborative networks and build global capacity among researchers and ultimately in the workforce. In the area of CCUS, bi- and multi-lateral mechanisms for collaboration, such as the US-China Clean Energy Research Center's Advanced Coal Technology Consortium and the Carbon Sequestration Leadership Forum, have already been established. These efforts together create a globally integrated collaborative network. However, US-China collaboration stands out among international efforts on CCUS because of the sustained nature of the collaboration and the level of commitment among the businesses and researchers engaged.

US BENEFITS FROM US-CHINA CLEAN ENERGY COOPERATION

There was a time when US-China collaboration on clean energy was geared primarily towards "capacity building" or "technology transfer." That view of the world is increasingly outdated. China is a global leader in clean energy investment and in clean energy technology development and the benefits from US-China collaboration on clean energy can and will be realized by both countries. China is now a center of global clean energy collaboration, working with Australia, Canada, and European Union on a range of clean technologies.

U.S. businesses, universities and think tanks are already leveraging the benefits of international collaboration to advance clean energy technology and reduce the environmental impacts associated with continued fossil fuel use.

In a world where companies and products are globally integrated, the benefits of US-China cooperation on clean energy technology innovation extends beyond either the US or China. A visit to an industrial facility anywhere the world might reveal parts and technologies purchased from dozens of countries. By leveraging and combining the collective ingenuity of engineers and scientists in both the US and China, we can help unlock a clean energy revolution by developing smarter approaches to producing and using energy. These new developments can and will be used globally.

An intangible benefit of cooperation is the benefit of the cooperation itself. And the benefits of this cooperation are not merely bilateral. The frameworks under which cooperation occurs provide opportunities for businesses, government agencies, and academics within each country to engage in discussions and work jointly in ways would not be possible otherwise. Such cross-fertilization directly benefits the US entities involved in the collaboration.

Before I turn to my recommendations, I would like to end my testimony with a concrete example of how collaboration under Clean Energy Research Center-Advanced Coal Technology Consortium and the US-China Energy

Cooperation Partnerships has resulted in deployment of new coal technology and created manufacturing jobs in the United States.[9]

LP Amina, a North Carolina-based company, developed and patented a new coal classifier to sort pulverized coal. The classifier was easy to add to existing plants and such a retrofit would lead to a reduction in coal consumption and emissions. The classifier prevents larger coal particles from entering the boiler, reducing nitrogen oxide emissions by up to approximately 15 percent and with slight efficiency improvements. Despite the benefits, customers in the United States would not buy the new classifier because it had not yet been demonstrated. After engagement in joint R&D and workshops convened by the CERC-ACTC, LP Amina installed one of its new classifiers at the Fengtai Power Station in the Anhui Province in eastern China. After the successful demonstration in China, LP Amina is now marketing this technology to global companies, including plants in the United States. The technology was developed here in the United States, demonstrated in China, and is now being deployed globally. And the innovation is creating American manufacturing jobs. Each classifier keeps 10-20 manufacturing workers busy for a month and manufacturers in Michigan, Ohio, and West Virginia have already been put to work building them.

RECOMMENDATIONS FOR IMPROVING US-CHINA COOPERATION ON CLEAN ENERGY

Note: the final three recommendations are an adapted and expanded version of the recommendations made for US-China collaboration on shale gas in "The United States and China: Moving towards Responsible Shale Gas Development"[10].

1. Creation of collaborative networks that include supporting innovators' needs and building capacity in the workforce.

While deployment of clean energy technologies depends on national regulatory changes and infrastructure investments, innovation also can be accelerated by collaborative networks. US-China collaboration that fosters true innovation and is sustained over time can play an important role in clean energy innovation. Such collaboration should extend beyond R&D and also build capacity in the workforce. The framework for such sustained, capacity-building collaboration is already in place through the US-China Clean Energy Research Centers, the Energy Cooperation Partnership Program and the US-China Renewable Energy Partnership—however, the budgets for these efforts are continually at risk of inadequate resources as they are subject to annual Congressional appropriations. In addition to ensuring collaborative efforts are adequately funded, Congress can also play a role in tracking the progress of various collaborative efforts through a periodic assessment or report to congress. Such a report could include recommendations for improving and reshaping existing collaborations.

2. Continued high-level strategic engagement between the US federal government and China's central government on clean energy collaboration.

High-level communication and engagement is very important. China's top-down innovation system requires government leadership and guidance. An institutional collaborative mechanism between China and the US could be beneficial. Such a platform should integrate current platforms and promote information sharing-among complimentary efforts. For example, although some of the same businesses and researchers are engaged in the shale gas initiatives led by different governmental agencies, there does not seem to be a formal integration or assessment of efforts. Ideally such an integration would look across clean energy cooperation, including energy efficiency, renewable energy, CCUS and shale gas. The Climate Change Working Group (CCWG) is designed explicitly to bring the relevant agencies and ministries together in such a manner. However, its scope is currently limited to collaboration on the five issue areas described. Congress could request a report that analyzes the gaps and progress in current collaboration and evaluates whether the CCWG could effectively serve a coordinating function among clean energy cooperation between the two countries.

3. Substantive work on developing and implementing environmental regulations for air, water, and climate impacts.

Too often collaboration focuses only on technology and not on the important interaction between technology and policy, including addressing the environmental impacts of technology deployment. Comprehensive, life cycle environmental concerns regarding clean energy development must be addressed as part of future collaboration. This collaboration should involve both technical and policy aspects of clean energy deployment. One way to accomplish this would be for the US and China to initiate a platform for multi-agency/ministry dialogue between the countries that is focused specifically on environmental policies needed for clean energy deployment in China and the US. Such a platform could be led by two agencies, or perhaps occur under the auspices of the CCWG, but would include all relevant agencies/ministries in the discussions and would focus on policies and regulations rather than technologies and demonstrations. Congress could encourage such collaboration through supporting international collaboration efforts undertaken by Environmental Protection Agency.

4. Private sector engagement and investment, personnel training, and more transparency.

Business-to-business collaboration will continue to be important, but there are barriers that must be addressed for it to be effective. They include inadequate involvement of companies on both sides, and sensitivities surrounding information protection and technology sharing.

To remedy these issues, the two governments could initiate a joint industry forum focused on the interests and needs of companies that cross-cuts clean energy issues. A new workforce training program could provide opportunities for young professionals from both countries to have an extended stay in the respective country. This would help to cultivate a group of engineers and scientists, ready to support clean energy development anywhere in the world. Leaders in both governments should jointly discuss the barriers to cooperation including intellectual property and data availability. Such an effort could begin by Congress requesting a special report on US-China business-to-business collaboration that includes input from industry, government, and academic stakeholders engaged in cooperative efforts and assessing the opportunities and barriers for private sector engagement in collaboration.

5. Facilitation of opportunities for joint R&D.

While existing technical collaboration is important, it should evolve into deeper efforts that include exchanges of researchers and dialogues aimed at solving the unique challenges of various clean energy approaches. Both governments should create platforms for cooperation and fund researchers in academia and industry to partner bilaterally and co-develop novel clean energy solutions.

Taking this recommendation seriously means prioritizing extended exchange programs over short workshops and meetings as part of existing and new collaborative efforts. It also means expanding beyond the questions around technology and including joint research on the economic and policy aspects of clean development. A collaborative clean energy demonstration program, modeled on existing international collaborations such as the CERC, could be one way to effectively encourage joint green technology development and demonstration.

End Notes

1 http://www.brookings.edu/~/media/Events/2014/2/06%20china%20clean%20energy/ USChina % 20Moving%20Toward%20Responsible%20Shale%20Gas%20Development_SForbes.pdf

2 These examples are excerpted from WRI China FAQS factsheet "US-China Collaboration: Can they "inspire the world"? (http://www.chinafaqs.org/library/chinafaqs-us-china-collabo ration-can-theyinspire-world)

3 Forbes, Sarah, 2012. "China's Prospects for Shale Gas and Implications for the United States" Congressional Testimony. http://pdf.wri.org/testimony/forbes_testimony_china_shale_gas_ 2012-01-26.pdf

4 http://www.uschinaogf.org/

5 http://www.fossil.energy.gov/international/InternationalPartners/China.html

6 http://www.rpsea.org/attachments/calendarevents/ 452/US-China%20Shale% 20Gas%20Res ource%20Characterization%20and%20Assessment%20Workshop-%20US%20V% 204_3_ 2013)USE.pdf

7 http://www.wri.org/publication/delivering-on-the-clean-energy-economy

8 http://pdf.wri.org/factsheets/factsheetpowerofinnovation.pdf and http://www.wri.org/ public cation/two-degrees-of-innovation

9 WRI published a more comprehensive version of this story as a ChinaFAQs factsheet; see http://www.chinafaqs.org/files/chinainfo/China%20FAQsRoadTestingTechv16.pdf.

10 http://www.brookings.edu/~/media/events/2014/2/06%20china%20clean%20energy/uschina% 20moving%20toward%20responsible%20shale%20gas%20development_sforbes.pdf

INDEX

#

21st century, 45

A

access, 16, 17, 18, 21, 86, 87, 89, 108
accounting, 6, 64
actual output, 65
AD, 16, 17, 66, 75
adjustment, 23
adverse weather, 36
Africa, 69
agencies, 68, 85, 87, 93, 100, 102, 109, 110, 112, 114, 115
air emissions, 92
air quality, 5, 47, 88, 90
alternative energy, vii, 2, 4
American Recovery and Reinvestment Act, 13, 57, 73
American Recovery and Reinvestment Act of 2009, 13
antidumping, 16, 17, 65, 74
appropriations, 114
arbitration, 28
Asia, 9, 45, 47, 70, 76, 77, 82, 97
assessment, 33, 114
assets, 22, 23, 27, 28, 86, 108
asymmetry, 108
authorities, 47
automation, 34

B

bankruptcy, 27, 77
barriers, 17, 115
base, 12
basic research, 99
behaviors, 34
Beijing, vii, 1, 23, 28, 45, 46, 47, 56, 67, 68, 72, 73, 75, 76, 77, 82, 88, 95, 108
Belgium, 76, 77
benefits, 5, 48, 88, 90, 98, 104, 107, 112, 113
Bilateral, 65, 98
biofuel, 107
biomass, 65
blogs, 77, 81
bondholders, 27
bonds, 28, 78
Brazil, 9, 31, 41, 75
business model, 9
businesses, 34, 86, 88, 112, 114

C

cadmium, 36
candidates, 43
capacity building, 112

carbon, vii, 1, 5, 9, 10, 25, 36, 41, 45, 46, 48, 64, 71, 79, 80, 83, 90, 101, 107, 110
carbon emissions, vii, 1, 5, 45, 46, 47, 48, 71
casting, 36
cation, 116
Census, 15
challenges, 4, 10, 15, 72, 85, 87, 88, 94, 98, 103, 115
Chamber of Commerce, 89
chemicals, 45, 70
Chicago, 28
Chinese firms, 17, 26, 36
Chinese government, 3, 4, 15, 16, 19, 22, 25, 26, 28, 34, 42, 46, 47, 48, 87, 88, 90, 94, 100, 110
cities, 3, 22, 34, 84, 90, 91, 106, 109
civil society, 105
clean air, 90, 100
Clean Air Act, 47
clean energy, vii, 2, 5, 6, 13, 15, 17, 29, 33, 48, 65, 85, 93, 94, 97, 98, 99, 100, 101, 103, 104, 105, 106, 107, 108, 110, 111, 112, 113, 114, 115, 116
clean technology, 6, 47, 88
climate, vii, 1, 24, 46, 48, 77, 88, 97, 98, 99, 101, 105, 106, 114
climate change, vii, 1, 48, 88, 98, 99, 101, 106
climate change issues, 99
CO_2, 46, 90
coal, 4, 8, 10, 11, 12, 22, 23, 34, 40, 41, 42, 43, 44, 45, 69, 98, 99, 100, 102, 106, 109, 113
collaboration, 88, 89, 93, 94, 99, 104, 106, 107, 108, 109, 110, 111, 112, 113, 114, 115
command economy, 22
commerce, 75
commercial, 14, 28, 64, 86, 87, 91, 92, 101, 103, 109
communication, 92, 114
community(s), 88, 99, 111
comparative advantage, 3
competition, 5, 9, 27, 48, 67
competitiveness, 9, 32
competitors, 15
complexity, 36
compliance, 4, 32, 70
composites, 36
conference, 91
confidentiality, 102
Congress, 2, 13, 14, 47, 64, 70, 84, 85, 114, 115
connectivity, 3, 11, 12, 20, 21, 22, 24, 33, 65
consensus, 45, 65
conservation, 100
consolidation, 78
constant rate, 65
construction, 21, 28, 34, 43, 82, 90
consumers, 3, 5, 15, 17, 22, 34, 88, 99
consumption, 32, 38, 39, 41, 42, 48, 69, 113
contamination, 42
cooperation, vii, 2, 4, 5, 9, 47, 71, 85, 87, 88, 90, 92, 93, 97, 98, 99, 100, 101, 102, 103, 104, 105, 106, 112, 114, 115
copper, 79
copyright, 28
cost, 3, 5, 8, 9, 14, 15, 23, 28, 32, 36, 45, 48, 64, 65, 111
countervailing duty, 16
covering, 26, 100
crises, 64
cross-fertilization, 112
crude oil, 39
crust, 70
crystalline, 31, 35, 36
Cuba, 28, 78
currency, vii, 1
customers, 23, 30, 113
CVD, 16, 17, 66

D

damages, 28, 68
data availability, 115
deficit, 29
delegates, 92
demonstrations, 89, 109, 115

Denmark, 12, 31, 73
Department of Commerce, 90, 94
Department of Energy, 6, 71, 72, 73, 74, 79, 80, 100, 101, 102, 110
Department of Justice, 28
Department of the Interior, 14
deployments, 109
deposits, 42, 70
depth, 110
designers, 30, 68
developed countries, 48
developing countries, 25
dialogues, 115
directives, 23
disaster, 43
disposable income, 48
distribution, 22, 34, 42, 66, 91
District of Columbia, 66
division of labor, 30
DOC, 75
domestic economy, 86
dominance, 10, 23
drawing, 106
dumping, 2, 16, 17
durability, 36
dysprosium, 32

E

early warning, 90
earnings, 28
earthquakes, 42, 44
East Asia, 24, 90
economic development, 24, 86, 87, 98, 105
economic downturn, 39
economic growth, vii, 1, 4, 31
economics, 83
economies of scale, 30
ecosystem, 111
electricity, 2, 3, 5, 8, 10, 11, 12, 14, 16, 20, 22, 23, 24, 34, 36, 39, 41, 42, 64, 65, 66, 69, 72, 73, 74, 79, 91, 98, 100, 111
electricity bills, 3
electricity consumers, 3
e-mail, 72, 73, 75, 82, 83, 84
embassy, 47
emergency, 90
emergency response, 90
emerging markets, 3, 28, 85, 86
emission, 46, 84, 90
emitters, 45, 99
employees, 28
endowments, 38
energy consumption, vii, 1, 22, 25, 39, 46, 90
energy efficiency, 41, 88, 89, 90, 99, 100, 101, 107, 109, 114
energy prices, 24, 31, 38, 41
energy production, vii, 2, 24, 107, 108
energy security, vii, 1, 4, 5, 38, 41, 47, 69, 88, 98, 109
energy supply, 44, 101
enforcement, 47
engineering, 25
Enhanced Oil Recovery, 110
environment(s), 5, 32, 47, 48, 75, 77, 82, 84, 88, 90, 100, 105, 111
environmental impact, 70, 89, 90, 112, 114
environmental issues, 89
Environmental Protection Agency (EPA), 70, 71, 90, 93, 100, 115
environmental regulations, 106, 114
equipment, 2, 5, 9, 10, 15, 16, 17, 18, 19, 21, 24, 27, 30, 36, 48, 65, 93, 111
equity, 67, 69
Europe, 14, 15, 30, 31, 46, 73, 99
European Commission, 16, 65
European Union (EU), 2, 4, 14, 32, 65, 112
evidence, 17
evolution, 71, 72
exchange rate, 23, 33
expertise, 85, 87, 93
export subsidies, 16
exporters, 31, 32
exports, 2, 17, 18, 19, 30, 31, 36, 86, 93, 111
exposure, 88
external constraints, 47
externalities, 4
extraction, 4, 67, 108

F

factories, 2, 9, 22
faith, 102
farms, 3, 10, 12, 18, 74
federal government, 106, 114
feedstock, 31
fiber, 36
films, 80
financial, 23, 26, 27, 64, 69, 108
financial crisis, 23, 26
flexibility, 3, 87
flight, 107
force, 12
foreign assistance, 86, 87, 93
foreign companies, 108
foreign exchange, 26
formation, 94
formula, 22
France, 36
fuel cell, 92
fuel consumption, 92
funding, 14, 16, 86
funds, 16, 24, 74, 75, 102

G

gallium, 36
GDP, vii, 1, 4, 13, 39, 40, 46, 90
geology, 4
Georgia, 43
Germany, 6, 7, 9, 10, 19, 36, 43, 80, 82
GHG, 109
global leaders, 19
goods and services, 9, 16, 64, 88
governments, 13, 18, 24, 43, 99, 105, 110, 111, 115
grant programs, 16
grants, 86
green buildings, 109
greenhouse, vii, 1, 46, 88, 92, 98, 99, 101, 109
greenhouse gas(s), vii, 1, 46, 88, 92, 98, 99, 101, 109
greenhouse gas emissions, vii, 1, 88, 101, 109
grid connectivity, 3, 11, 20, 21, 24, 65
grid technology, 3, 34
grids, 14, 20, 34, 79, 107
gross domestic product, vii, 1
groundwater, 42
growth, 4, 19, 22, 24, 31, 39, 46, 77, 108
growth rate, 19
Guangdong, 16, 74
guidance, 114

H

heavy metals, 90
history, 78, 86, 99, 103
holding company, 27
homes, 34
Hong Kong, 27, 60, 78
host, 90, 91
House, 73
housing, 39
hub, 31, 35, 36
human, 36
human health, 36
hybrid, 36
hydrofluorocarbon emissions, vii, 1

I

identification, 98
image, 83
imbalances, 19
imports, vii, 1, 5, 15, 17, 18, 31, 32, 38, 39, 42, 65, 69
improvements, 9, 25, 28, 35, 39, 104, 113
income, 16, 28, 41, 48, 85
incumbents, 9
independence, 106
India, 7, 10, 27, 41, 70, 75, 82
indium, 36
industrial policy, 4, 18, 26, 48, 65
industry(s), vii, 1, 2, 3, 5, 6, 9, 13, 21, 26, 27, 30, 31, 36, 39, 41, 42, 44, 47, 65, 69,

86, 87, 88, 89, 90, 91, 93, 101, 107, 108, 109, 110, 111, 115, 116
industry consolidation, 26, 27
inequality, 41
infancy, 34
inflation, 31, 41
information sharing, 114
infrastructure, 14, 21, 24, 34, 65, 85, 86, 87, 88, 111, 113
insecurity, 39
insertion, 9
inspectors, 109
institutions, 87
integration, 3, 22, 27, 34, 72, 114
intellectual property, 9, 17, 28, 101, 102, 104, 115
intellectual property rights, 104
interference, 42
International Energy Agency, 19, 46, 69, 82
international law, 103
international standards, 39
inventions, 103
investment(s), 2, 6, 11, 13, 14, 15, 17, 26, 34, 48, 88, 91, 106, 108, 111, 112, 113, 115
investors, 13, 43, 82
invitation to participate, 97
IPO, 67
iron, 36
irrigation, 42
isolation, 12
issues, vii, 2, 15, 20, 21, 48, 65, 83, 89, 98, 100, 101, 104, 106, 109, 110, 115

J

Japan, 4, 10, 23, 27, 32, 43, 99
job creation, 87
joint ventures, 108
jurisdiction, 15

K

Kyoto Protocol, 45

L

language policy, 41
Latin America, 28
laws, 103
lawyers, 102
lead, 3, 6, 19, 23, 87, 98, 102, 104, 113
leadership, 91, 102, 114
learning, 108
legislation, vii, 1
liberalization, 4
life cycle, 114
lifetime, 48
light, 36, 41, 103
liquefied natural gas, 42
litigation, 79
loan guarantees, 13
loans, 16
local government, 24, 41
longevity, 111

M

magnet, 28, 32
magnets, 32, 36
magnitude, 70
majority, 6, 18, 34, 67
management, 90, 91
manufacturing, vii, 1, 2, 9, 13, 15, 18, 20, 23, 26, 30, 31, 32, 36, 47, 48, 80, 113
mapping, 109
market access, 17
market position, 26
market share, 4, 9, 31, 35, 67
marketing, 113
marketplace, 111
materials, 30, 32, 36
matter, 47, 90
measurements, 25
media, vii, 2, 47, 70, 84, 116
medical, 88
mercantilism, 16
mercury, 90, 94
metals, 32, 36

meter, 34, 39, 67, 89
methodology, 110
Middle East, 69
minimum price, 66
mission(s), 86, 89, 90, 91, 92
models, 97
modifications, 35
modules, 9, 15, 17, 18, 19, 35, 65, 75
molybdenum, 79
monopoly, 22, 32
Montreal Protocol, 45
motivation, 24
multilateralism, 98
multimedia, 80
multiples, 19

N

nanotube, 36
nanotube films, 36
national borders, 9
National Party, 47
National Renewable Energy Laboratory (NREL), 36, 73, 80
natural gas, vii, 2, 11, 39, 41, 42, 44, 106
natural resources, 32
Nd, 32
negotiation, 98
neodymium, 32
nitrogen, 90, 113
North America, 42
NREL, 36, 72
Nuclear Regulatory Commission, 43

O

Obama, 17, 45, 100
Obama Administration, 17, 45, 100
obstacles, 97
Office of the United States Trade Representative, 75
officials, 3, 4, 22, 24, 40, 41, 71, 109
offshore wind capacity, 2, 14
oil, vii, 1, 4, 38, 39, 40, 41, 42, 64, 69, 70, 81, 106, 108, 110
oil production, 4, 69, 70
onshore wind, 3, 8, 13, 21
operations, 33, 36, 64
opportunities, 9, 85, 86, 87, 88, 90, 91, 93, 106, 109, 112, 115
optimization, 91
outsourcing, 36
overproduction, 34
overseas investment, 26
oversight, 102
ownership, 23, 39, 102
ozone, 45

P

Pacific, 45, 47
Pakistan, 16, 28, 74
Panama, 28, 78
parallel, 20, 22
parity, 34
participants, 92, 102, 103, 104, 110
petroleum, 11, 31, 32, 41, 42, 46, 69, 83, 108, 110
pipeline, 43, 69
plant type, 64
plants, 8, 43, 45, 65, 113
platform, 5, 89, 109, 114, 115
playing, 9, 14, 47, 92, 102
policy, vii, 1, 2, 18, 19, 20, 26, 66, 87, 89, 97, 99, 101, 102, 114, 116
policymakers, 24, 42, 111
political leaders, 44
pollution, 17, 47, 69, 84, 88, 90
population, 4, 34, 91, 98
portfolio, 13, 18, 66, 90
Portugal, 78
potential output, 65
power generation, 2, 3, 9, 10, 11, 12, 13, 16, 22, 24, 26, 27, 43, 67, 69, 91
power lines, 20, 69
power plants, 41, 44, 45, 90, 107
premature death, 88
president, 5, 15, 25, 45, 83, 88, 109

President Obama, 45, 88, 109
private investment, 13
private sector, 85, 87, 93, 103, 111, 115
procurement, 41, 86
producers, 15, 16, 17, 23, 26, 27, 31, 36, 79
production costs, 66
professionals, 115
profit, 23, 28, 78, 105
project, 14, 16, 28, 34, 36, 48, 67, 69, 74, 75, 78, 85, 86, 87, 89, 90, 91, 92, 93, 97, 101, 102, 103, 110
project sponsors, 93
protection, 36, 115
public health, 5
public investment, 111
public sector, 13, 87
public-private partnerships, 85, 89, 93
purification, 36

Q

quality control, 65
quality of life, 48
quotas, 32, 79

R

race, 64
radiation, 36
ramp, 31
rare earth elements, 4
RDD, 102
reading, 88
real income, 41
real wage, 31
recommendations, 106, 112, 113, 114
refinery capacity, 69
reform(s), 22, 23, 65, 73, 75, 81, 82, 100
regulations, 13, 68, 92, 115
regulatory agencies, 65
regulatory changes, 113
regulatory framework, 91
reliability, 92
renewable energy, vii, 1, 2, 3, 5, 9, 10, 13, 15, 16, 18, 19, 20, 22, 23, 24, 34, 41, 48, 65, 66, 91, 99, 100, 107, 109, 114
requirements, 4, 36, 85, 92
researchers, 101, 102, 103, 104, 109, 110, 111, 114, 115
reserves, 4, 24, 26, 42, 69
resolution, 103
resources, 15, 24, 25, 26, 27, 47, 50, 67, 70, 72, 81, 82, 87, 89, 98, 106, 114
response, 87, 89, 92
restrictions, 4, 32, 36
restructuring, 27
retail, 22
retaliation, 17
revenue, 27, 28, 41
rights, 27
risk(s), 43, 48, 69, 70, 82, 108, 114
Romania, 28, 78
rules, 17, 32, 102, 103
Russia, 28, 78

S

safety, 5, 41, 43, 44
savings, 21, 40, 91
scarcity, 42
science, 102
scope, 107, 114
second generation, 64
security(s), 38, 68, 100
SED, 100
September 11, 72, 75
service provider, 108
services, 3, 10, 27, 47, 48, 64, 68, 86, 107, 108
shareholders, 28, 67
signs, 25, 74, 78
silicon, 4, 31, 35, 36, 48
smog, 84
social costs, 4
solar cells, 15, 17, 36
solar power, 2, 3, 6, 11, 15, 16, 20, 34, 38, 45, 47
solar PV equipment, 2

solution, 4, 41
South Korea, 43
spending, 13, 111
spin, 36
stability, 24
stakeholders, 30, 93, 94, 115
state(s), 2, 4, 6, 8, 12, 16, 22, 23, 24, 25, 28, 34, 42, 43, 47, 66, 69, 70, 71, 73, 75, 88, 92, 95, 106, 110
state-owned banks, 16
state-owned enterprises, 22
statistics, 31
steel, 32
stock, 69, 109
storage, 20, 34, 107, 109
Strategic Economic Dialogue, 100
stress, 42
structure, 88
style, 89
subsidy(s), 57, 65
sulfur, 90
sulfur dioxide, 90
Sun, 20, 66
superconductor, 79
suppliers, 3, 19, 24, 28, 30, 34, 36, 43, 48, 53
supply chain, 2, 3, 5, 36, 47, 48, 90
surface area, 36
surplus, 30, 31
susceptibility, 44

T

Taiwan, 9, 17
tanks, 112
target, 18, 20, 25, 31, 46, 66, 76, 90
tariff, 18, 21, 22, 76
tax credits, 13, 14
tax incentive, 13
tax reform, 4
taxes, 22, 41
taxpayers, 2
teams, 100
technical assistance, 86
techniques, 108
technological advances, 3
technological developments, 5
technology(s), 2, 3, 4, 5, 6, 14, 17, 26, 28, 33, 34, 35, 36, 42, 43, 44, 48, 68, 78, 80, 85, 87, 90, 91, 92, 97, 99, 100, 101, 102, 103, 107, 108, 109, 110, 112, 113, 114, 115, 116
technology transfer, 17, 97, 112
telecommunications, 87
tellurium, 36
tensions, 42
territory, 25, 102
testing, 36
theft, 17, 28
thermal power, 3, 11, 23, 101
Tibet, 67
top-down, 114
tracks, 41, 102
trade, 2, 3, 5, 17, 19, 28, 29, 31, 46, 47, 70, 86, 87, 89, 90, 91, 92, 94, 111
trade deficit, 29, 111
training, 106, 109, 110, 115
trajectory, 23
translation, 75
transmission, 4, 11, 21, 22, 24, 33, 69, 91, 98, 100
transparency, 106, 115
transport, 34, 36, 39, 42, 69
transportation, 87, 92, 100
treatment, 16, 69
trial, 28
tungsten, 79
turnover, 28

U

U.S. Department of Commerce, 16
U.S. Department of the Interior, 73
U.S. policy, 101
U.S. Treasury, 100
unit cost, 8
United Kingdom (UK), 9, 10, 14, 64, 77
United Nations (UN), 5, 25, 46, 48, 98, 103
United Nations Framework Convention on Climate Change, 98

universities, 46, 99, 103, 111, 112
urban, 21
urbanization, 39
USA, 28, 78
utility companies, 2, 14, 18, 23

V

vehicles, 39, 45, 92, 100, 109
vested interests, 3
vision, 110
volatility, 38, 69

W

wages, 31
Washington, 30, 45, 69, 71, 73, 74, 77, 78, 79, 80, 81, 82
waste, 42, 65, 88
waste management, 65
waste water, 42
water, 4, 42, 69, 70, 81, 82, 90, 100, 106, 114
water resources, 42
water supplies, 69
wavelengths, 36
wealth, 5, 26
web, 73, 80
wells, 70, 108
White House, 71, 77, 83, 94
White Paper, 76
wholesale, 22
wind farm, 3, 8, 11, 12, 13, 14, 15, 16, 18, 19, 20, 21, 24, 26, 27, 28, 36, 68
wind power, 6, 8, 11, 12, 14, 16, 18, 21, 23, 25, 27, 31, 48, 66, 67
wind speed(s), 12, 21, 36
wind turbines, 8, 18, 31, 35, 67
workers, 113
workforce, 106, 111, 113, 115
working groups, 89
working hours, 33
World Bank, 81
World Health Organization, 88
World Trade Organization (WTO), 4, 17, 32, 47, 75
worldwide, 6, 9, 14

Y

Yale University, 80